GREEN GUIDES

Keeping Chickens

LIZ WRIGHT

Foreword by TRACEY SMITH

FLAME TREE

Contents

Aside from providing fresh eggs, chickens are friendly, intelligent animals with unique personalities, so they make great companions. There are a wide variety of breeds and it is important to do research for what breed best suits your own goals. Chickens can help out by eating pests in your garden, they can be an educational entertainment for children, or be show animals for hobbyists. Keeping your own chickens is an environmentally friendly and rewarding decision.

From space concerns to legal issues, there are many things to consider before deciding that keeping chickens is right for you, and, once you have decided, you must prepare for their arrival. In order to care for chickens, you must be sure to protect them from local predators and hazards. There are several housing options for your fowl and you should consider the advantages and disadvantages of each one. This chapter provides everything you need to know about the materials and equipment you need to obtain in advance of getting your chickens.

Buying Your Chickens

Now that you have decided to buy chickens, you are faced with new decisions such as which breed to get and whether or not it is appropriate to buy a cockerel. This chapter offers information for considering these questions as well as offering advice on the size of the breed and what age the chickens you are buying should be. Whether you decide on hybrid or pure breed, this chapter will inform you on how many and what type of eggs to expect and any particular considerations for each breed. It also includes pictures of every breed you might consider buying.

Feeding Your Chickens

In order to keep your chickens happy and healthy, it is necessary to understand their digestive system as well as how and what to feed them. Laying eggs is a demanding process and it is important to provide the hens with all the nutrition required for the task. This chapter also provides details on the different types of food, how to store it, how often chickens should be fed and which type of feeders and drinkers are right for your birds.

Caring for Your Chickens

Once you have your chickens, you must observe their behaviour. Each flock will develop a strict pecking order and introducing new chickens can have consequences – sometimes fatal. To be a responsible caretaker, you must be able to handle them properly, clip their wings if necessary and be on the lookout for illness. This chapter provides checklists of routine tasks necessary for keeping your chickens healthy and safe. In addition, it gives symptoms of and potential treatments for common poultry health problems as well as providing information on vaccinations and veterinarians.

Reaping Your Rewards

Now that your hens are housed and cared for, they will start to lay eggs. In order to benefit from this new income, you must know how and when to collect your eggs and how to store them. Generally, eggs are clean, but in some cases it is necessary to cook the egg as thoroughly as possible to avoid bacteria. If you are rearing table birds, they should be slaughtered humanely and prepared properly before cooking. This chapter also provides tasty recipes featuring eggs and poultry.

Pests, Predators & Problems

From red mites to foxes, there are a wide variety of parasites and predators to watch out for in order to keep your chickens safe and healthy. In addition, though some behaviours are natural and not something to worry about unduly, there are many unusual behaviours that can prevent your poultry from thriving as they should. For example, chickens may begin eating their own eggs or pecking at others in the flock. It is important to know how to deal with these problems when keeping chickens.

Adding to Your Chickens

If you have found your experience to be rewarding, you may want to add new chickens to your flock. Introduce new chickens in a way that will cause minimal disruption to the pecking order. If you plan to breed your chickens, it is important to understand the reproductive process and to decide whether to hatch by hen brooding or artificial incubation. Once the chicks have hatched, special care should be taken of them, because they are more delicate than their full-grown counterparts.

Foreword

When we first toyed with the idea of keeping chickens, I must admit, the only benefit I could see was that we would have fabulous free-range eggies; how wrong could I have been? Well, very wrong, to tell you the truth... Being bountiful in eggs was just the icing on top of the cake, made with said eggs, of course.

We did our research and decided to start off with 10 point-of-lay, unexotic, uncomplicated dual-purpose birds, which we bought from a reputable source at a local market. We liberated the youngsters into a rickety but well-sealed old chicken house within a large enclosed run that had upstairs and downstairs accommodation. We placed six wooden wine crates on the top floor in a tidy row and filled them with hay, hoping the chooks would suss out that that was where we'd like them to lay their precious cargo.

All credit to them, they did. The only trouble was, they all seemed to favour the box on the far right-hand side and would often queue up and sometimes clamber on top of each other to get dibs on that particular crate, which to an onlooker was utterly perplexing! Their run also housed a small tree with low branches and the girls used to hop up into it for a daytime roost.

I quickly realized there was far more to these endearing creatures than I'd anticipated, so readily plopped two garden chairs in their run so my husband Ray and I could sit and relax with them while we shared our morning cuppa and chewed the cud.

The girls thoroughly enjoyed their own company, but also revelled as a pack and, in the height of the midday sun, would lay together in the shade of the tree, one using a fellow chook's wing as her pillow. Dust-bathing on a balmy afternoon was always a sight to behold and there was a well-used dip near the blackberry bush that became 'the' spot to do it.

As a result of observing their predictable daily routines, it was clear to see their expectations from the world could be totted up on the digits of one hand; clean water, food, company and a safe shelter, oh, and the odd foraged bug, made their lives complete.

There were many lessons to be gleaned from their straightforward approach to life and they helped me analyse and re-prioritize the madness that is modern-day living. They also helped lower my pulse rate.

The addition of 10 one-year-old ex-battery hens to our flock was positively life changing. Hot tears burned scars on to my cheeks as I placed them on soil for the first time ever and, within moments, their God-given instinct to scratch the earth had kicked in. They had been de-beaked, had very long claws and barely any feathers, but within weeks they had made a remarkable recovery and were very much part of the gang, and when my children approached them to collect their eggs, they would be amongst the first to squat down, spreading their wings in a 'please tickle me' position, which was heart-warming to see.

Muscovy ducks (10 of them) and two guinea fowl completed our full set of feathered friends and they were an absolute joy to look after, but none were ever quite so lovable as the chickens, who lived to a ripe old age and continued to bless us with eggs for many years.

Chickens: there's so much more to them than it says on the tin and this book is the perfect place to get started on your journey to finding that out!

Tracey Smith
Author and broadcaster on sustainable living.

Introduction

When I told a passing acquaintance recently that I was writing a book on keeping chickens, she said, 'Ooh! My best friend used to keep chickens. I always felt it slightly odd, but she loved it.' I mention it here because I've been saying for years that there are two kinds of people in this world: those who have experienced the joy of collecting a still-warm egg from bustling, busy birds and those who have yet to do so.

Are Poultry Different to Other Pets?

Dogs, cats, hamsters, goldfish, tortoises and ponies don't provide you with lovely breakfasts and dinners. They do have their particular benefits, but none of them relate to anything so primal and essential as food. Furthermore, keeping chickens is something people have done all over the world for many thousands of years and for the very same reasons that people keep them today. But they also become part of the family to a greater or lesser extent. Some people like to name their girls and encourage them to be tame, while others prefer to look after their flock but treat them in a more distant manner. Either way, they are still part of your daily life, but a part that produces fresh eggs with a financial as well as healthy value.

Where Can Poultry be Kept?

As long as their management requirements are met, chickens can be kept in a wide range of situations, from urban back yards to the countryside.

With over 100 pure poultry breeds, from bantams to large fowl, plus an ever-increasing choice of attractive and productive hybrids, there is a bird suitable for every type of location. There are even bantams being kept in the centre of New York! This book will help you to choose the right birds for you and guide you through the best way of keeping them so that you and your birds can enjoy a productive relationship.

The Chicken Revolution

More and more people in the modern world are waking up (metaphorically and literally) to the joys of keeping chickens. This recent rise in interest is part of a greater cultural movement that is pushing towards a calmer, healthier, cleaner, kinder and more sustainable way of living. The fact that this very book is a 'Green Guide' and is written by someone who believes in living at least some part of your life in a self-sufficient way, is testament to that fact.

All Welcome

Of course, you don't have to be an eco-activist to keep chickens. You might simply want them for the lovely pets they make, for their delicious eggs or meat, to proudly enter them into shows, to appreciate the beauty of their plumage or to be soothed by the sound of their gentle voices.

Sustainability and Chickens

Reasons for why keeping chickens has become so intimately associated with the movement towards greater sustainability are:

Food Control

Like allotments and home-grown food, they are a step towards taking back control of our food from mass industrialized agriculture, which is generally seen as unsustainable. More control also means that the keeper knows exactly how his eggs or meat are produced, what the birds were fed and how they were kept.

Natural Connection

Just like people who grow their own food, people who keep chickens develop a deeper understanding of their own food. They bring the keeper into a closer relationship and understanding of nature generally. Seasons, weather, bacteria, water, parasites, predators, bugs, damp, humidity and territory, amongst many other processes and phenomena, are all aspects of nature that chicken keepers come to know intimately through their chickens. Knowledge of natural processes, particularly in regard to food production, is a core element of the sustainability ethos.

Health and Pleasure

As with allotments and home-grown fruit and vegetables, chickens help to cultivate a greater respect and appreciation for food. The enjoyment of food, particularly in the relaxed company of family and friends, is a key feature of the green lifestyle, as it is seen to be better for health and happiness, both personal and social.

Top Tip

To help save the world and enjoy a happier life, get some chickens.

Local versus Global

More chickens in our villages, towns and cities means an increase in 'local' produce. Localization of food and other resources is a primary objective for the green movement. This is because it helps to strengthen the economies of local communities and reduces the damaging emissions of excessive transportation, not to mention the social justice issues associated with the global food industry.

Cruelty-free

More chickens in the community means fewer chickens in battery farms. People who actively support a greener lifestyle tend to have a profound abhorrence of cruelty and a strong inclination to take action on issues that matter to them. Keeping chickens is one way they can do something to actively help to reduce the cruelty of battery chicken farms.

The Good Life

I think that the very presence of chickens, their cautious movements, slow pace and soothing sounds, has an emotional and psychological effect that somehow encapsulates the very spirit of 'The Good Life', an idea as ancient as Aristotle but championed since the 1960s by the green movement.

What Do Chickens Need?

At the most basic level, chickens need housing, shelter, protection from predators, correct feeding and clean water plus attention to any illnesses or injuries that they might sustain. They also need the ability to express their natural behaviour such as scratching, wing flapping, foraging, pecking for food and movement. So your management is based upon meeting these basic needs and, in a back-yard situation where there are only a few chickens, it should also be the aim to go beyond this and provide even better care based on daily observation. In addition, the tamer the flock, the less stressful it is for them to be handled for inspection or for any treatment.

What Do I Need to Keep Chickens?

Time

Chicken keeping, like any livestock, requires a regular, routine commitment of time. Although this is not a large amount of time, it has to be twice daily and if the keeper cannot provide this, then they will have to make arrangements for someone else to do so.

A Little Money

Financially, after the initial set-up costs, which will vary according to whether there are existing buildings that can be adapted for housing, the maintenance cost of keeping chickens is relatively low compared to a dog and certainly compared to a pony! Equipment such as feeders and drinkers will need to be purchased and replaced from time to time  and, like all pets, there are treats and luxuries available for them if you want to pay for them. Although chickens are naturally healthy, if well managed, there are also vets' costs to consider.

How Do I Begin?

This book is designed to guide you through getting started with chickens, producing eggs and keeping your chickens' health. Choosing the correct breed for you is one of the keys to success.

The book will also take you on to meat production, breeding your birds and considering where to go next – such as taking part in poultry shows or adding to your collection. Dealing with minor illnesses and preventing parasites are also skills you will need – you will find these in the health and troubleshooting sections.

About this book

Throughout this book, you will learn:

- **What you need to start keeping chickens.**
- **Which chicken breed is for you.**
- **How to get the most from your chickens.**
- **How to use good management to keep your chickens happy and healthy.**
- **What a balanced diet means in feeding your birds.**
- **How correct handling is good for you and your birds.**
- **How to recognize common diseases.**
- **How to prevent parasites by regular routine treatment.**
- **How to take a step further into hatching and rearing.**
- **About other opportunities for poultry keepers.**

The Joy of Keeping Chickens

A Bird for All Reasons

Mostly, when we think about keeping chickens ourselves, we think of getting fresh eggs on an almost daily basis. Yet there are other reasons for becoming a poultry keeper. Certain breeds can be raised for meat and surplus cockerels can be culled for the pot. Beautiful breeds can be exhibited at poultry shows and their resulting youngstock can be sold as pure breeds. But don't underestimate the joy of keeping them as pets! They are intelligent (yes, they are), amusing and rewarding creatures.

Poultry for Pleasure

Providing you establish the correct housing and fencing for your birds, and get a routine for looking after them, then poultry really are a pleasure to keep, rewarding you not only with eggs but also with their enthusiasm when they see you. I love it when my hens run across the yard towards me, their feathers flying like skirts and their necks outstretched with eagerness.

Birds With Personality

I know it's the food they like really, but it's still a pleasure to see them and be surrounded by their enquiring voices. They turn their heads to one side to look up at you and you know they are more than just egg layers. And each one has a different character. There's the shy one who hangs behind, the bold one who bustles away in front and the noisy one who has plenty to say. You'll get to know them all.

Productive Poultry

With so many poultry breeds and types, there is a big difference in the number of eggs they lay. But generally speaking, all breeds will produce eggs at some point. A hybrid hen will produce around 300 eggs a year, while very pretty Japanese bantam types might only manage 20 or so in a year. Within this range, there is a wide variance of egg performance, with many breeds well known in the past for laying, such as the Orpington, now averaging around 150 a year unless you are able to find a good laying strain of the breed. Choosing the right breed for your circumstances is very important.

Birds for Meat

Chickens also produce meat, either accidentally – in that you cannot keep too many cockerels and have to do something with the surplus ones – or on purpose, by buying the breeds of chickens that are known for being good table birds. Again, there is much variety between the breeds.

Did You Know?

The Red Jungle Fowl (*gallus gallus*), which originated in Asia, is widely thought to be the ancestor of our current domestic poultry.

Chickens Can be Good Gardeners

Chickens love to eat bugs and often these bugs love to eat your plants and veg. If you let the chickens scratch over your veg plot after harvest and through the winter, they will help you in your quest to grow, as they will take out many of the harmful bugs and slugs while fertilizing the garden with their manure. Also, when you clean out the poultry house, you can put their manure into a composting area and recycle it to make nutritious compost for your garden.

But Also Indiscriminate Gardeners

Unfortunately, chickens are also quite partial to tender young seedlings and picking off leaves with their sharp beaks, so their gardening activities need to be managed. If you let them into your garden when plants are young or seedlings are small, then their natural habit to scratch with their powerful little feet will send seeds scattering!

Top Tip

Take time to read up on keeping chickens now to save time later when looking after them. This way, you'll get the right house and fencing, the breed of chickens for you and will know what you need to do on a daily basis.

Chickens and Children

Children are fascinated by chickens and love to collect eggs. They make excellent pets for children, although, as with all pets, children must learn to respect them and to handle them safely. Some bantam breeds, such as Pekins or Old English Game, get particularly tame if they are handled a lot from chicks. They are small enough for quite young

children to hold and are appealing in their looks and personalities. But even the more commercial breeds, such as hybrids, are loved by kids; they like to stand and feed them, watching them scratch and squabble, and they love to put their hands into the nest box and fetch out a still-warm egg. It's good for education too.

Easy to Keep

Poultry are easy to look after as long as you have taken the time to find out what they need and how to provide it for them. Unlike larger animals, they do not require *lots* of time and energy to keep clean, just regular attention to removing droppings and feeding, which is straightforward if you select the right diet and feeders. As long as you have a secure house and good fencing to keep out predators, then your poultry keeping will be enjoyable.

Your Chickens Need Some of Your Time

You'll need to find time at least twice daily to let them into the run or free range and then to shut them up safely at dusk, check water and food and look at their health. Once a week, you'll need to clean them out and check the fencing and housing.

Myths About Keeping Chickens

People will tell you that it's not as easy as you think keeping poultry – that they know people who had a mouse problem after they starting keeping chickens, that they spread disease or that neighbours complained about the noise levels, the loud crowing of the cockerel in the morning. Some people say that there are covenants on houses and estates that don't allow the keeping of chickens.

Am I Forbidden?

Firstly, check on your deeds if you think there might be a problem, but often this is just an urban myth. There is no general rule preventing the keeping of poultry in urban or suburban areas – sometimes there is a local rule, but years of answering questions from would-be poultry keepers have led me to believe that this is far less frequent than is thought.

Help! My Neighbours Are Worried About the Noise!

It is true that some breeds of poultry are more chatty than others, but happy poultry make a gentle contented noise except when announcing loudly that they have laid an egg. Even so, your neighbours would have to be very close to find this irritating.

No Need for Cockerels

What your neighbours might find annoying is being woken at daybreak by a cockerel. There is no need to keep a cockerel if you do not want to breed – chickens will lay eggs without a cockerel and it will not affect the number of eggs they lay. In fact, they may lay rather more without a cockerel. Hybrid hens in particular are happy without having a cockerel, as they have been bred to live in big laying flocks.

Top Tip

Site your poultry run away from the neighbour's fence and plant fast-growing shrubs alongside to deaden any noise.

Other Common Myths

Some problems that chickens are reputed to cause may be rooted in truth and others are complete myths – either way, they should be no barrier to your ambitions:

- **'Chickens attract vermin'**: Use suitable feeders, clean them regularly and store food in vermin-proof containers (an old dustbin or metal bin).

- **'Chickens are smelly and dirty'**: Only if you don't clean them out regularly. Take out dirty bedding and use in a managed compost bin.

- **'Chickens spread disease'**: Like most livestock, there are only a few things that can be passed on to humans, and keeping them clean and watching out for signs of ill health will reduce this risk to a minimal one.

- **'Chickens create problems with other pets'**: Pets are more likely to cause chickens problems if not kept under control. It would be common sense to keep small pets such as mice or hamsters away from chickens and it would be cruel to let dogs close to poultry, as the birds are scared of even well-meaning dogs. Cats usually pose no threat to adult birds.

- **'Chickens ruin gardens close by'**: Either clip flight feathers from wings (non painful) to stop them flying over fences, or keep them confined in a good-sized run.

Good for the Earth, Good for Us

Up until recently, 'food security' (or lack of) used to apply only to countries other than in the West, but now it's a term heard with more frequency in Europe and the USA, as the concerns about the sustainability of food grow. 'Food miles' were unheard of in the past when the only imported items were those we couldn't grow ourselves, such as bananas and oranges – making them luxuries to be enjoyed. Now almost everything we put in our supermarket basket has made an extensive journey. How can keeping chickens in your back yard help?

On Your Doorstep

The food 'miles' of your home-produced eggs will only amount to as many steps as it takes for you to go to the hen house and collect them newly laid. You could argue that their food will rack up some miles, but, if you use a local merchant, that will help. If you choose good layers and get the number needed to keep you in eggs all year round, then you need never have to drive to a shop for eggs and will never run out. You'll also know exactly when they were laid and how fresh they are.

Getting Started in Sustainability

Keeping chickens can lead you to create your own 'ecosystem', whereby their composted manure feeds your veggies and, together with a balanced bagged feed ration, your veg trimmings can be returned to your chooks. They will also have played their part in garden pest control, turning pesky plant eaters into nutrition for their eggs. But they can also lead to an even greater interest in self-sufficiency, such as pig keeping or planting orchards and using chickens to clear up waste fruit.

Good for Health and Wellbeing

Being outside and in touch with the seasons is good for your health and it is beneficial for children and adults to relate to living creatures. Also, the regular, gentle exercise in caring for your hens, plus a harder session of cleaning out, will benefit you physically.

Nutrition

Of course, you will also have a supply of fresh eggs for cooking. The benefits of eating eggs have now been recognized once more, with the concerns over increasing cholesterol levels now not thought to be a problem. Eggs are a good source of protein, as well as providing lots of other essential nutrients, such as selenium, iodine, molybdenum, phosphorus, vitamin B2, vitamin B5, vitamin B12 and vitamin D.

Did You Know?

Eggs imported for use in UK foods travelled more than 3.5 million miles in one recent year, the equivalent of travelling to the moon and back more than eight times, according to the British Egg Products Association.

Going Further Than Eggs

Commercial chicken meat has become a tasteless, low-fat, cheap white meat that lends itself to adding sauces, processing into pies and is ideal for fast foods. When you keep chickens of your own, you begin to realize that there is a chicken at the bottom of all this. The way they are kept for meat not only is an issue as far as their welfare is concerned, but the management and processing also impacts on the nutrition in your diet. So, although it's not for everyone, the next logical step is to keep some birds for meat.

'Accidental' Chicken Meat

Laying hens can be kept without a cockerel and they will still lay eggs, which for many will be the ideal method of management. But if you want to breed, you will need a cockerel. The watchword is 'a' cockerel, singular. As each clutch of eggs hatched usually contains about a

50:50 male to female ratio, it is very easy to end up with too many cockerels. This turns you into an 'accidental' chicken meat producer. You cannot find homes for all the cockerels and it is wrong to just dump them on a market, so suddenly you have meat chickens even though it's not what you intended!

Did You Know?

It is estimated that around 25–30 kg (55–66 lb) of chicken meat is eaten per person in the UK and the USA, almost doubling from the 1970s.

Purpose-raised Meat Chickens

As we will discover, laying and table birds are two entirely different things, apart from being chickens. A meat breed will often be a poor-to-moderate layer, will certainly have a good ratio of meat to bone and will be a weighty bird. There are traditional breeds of meat chicken such as the French Marans (now prized for its dark brown eggs) and the Jersey Giant (now prized for its 'Exhibition' qualities). But there are also purposely bred table birds that fatten at an alarmingly fast rate and are docile and slow moving. These are called broilers.

The Indian Game is often found in the breeding of a modern table bird.

Considerations for Chickens as Table Birds

- Learn the difference between laying breeds and table breeds
- Choose the correct feeds for their nutritional needs
- It is a responsibility to ensure slaughter is quick and humane
- You may need to learn to pluck and draw
- Purposely-bred table birds can be roasted, but older cockerels will need slower cooking or stewing

Showing off Your Chickens

Showing poultry can be a whole new avenue of enjoyment. Poultry shows first became popular around the late-nineteenth century as trade routes opened up and traders brought different breeds of poultry from around the world to Europe and the USA. In the twentieth century, they developed into shows that not only displayed the most exotic and exciting breeds of poultry, but also took farmyard fowl and utility fowl (suitable for table and a reasonable egg layer) and gave them classes too.

So, originally, a poultry farmer and breeder took a pride in exhibiting his commercial flock alongside the 'fancy' (more exotic) breeds of poultry. But, as the small poultry farmers started to disappear after the second world war, the shows changed again to become based more on 'Exhibition' qualities. By this time, specific 'Standards' had been written in the UK and USA for most breeds, and clubs for those exhibiting poultry had been formed.

A Cheap but Skilful Hobby

In comparison to most forms of hobbies, showing poultry requires a modest outlay. You need the best birds (the ones that meet the standards laid down for that breed as closely as possible), but after that, the entry fees to shows are modest and the other costs are just about how far you want to travel and investing in a suitable and humane poultry carrier. Participants often describe themselves as being 'in the Fancy', the 'Fancy' being a catch-all name for exhibition poultry.

Making a Start

First of all, visit a poultry show and look at the breeds of poultry. Talk to the exhibitors; they are only too happy to speak to novices and share their knowledge. Normally, hybrids and hens that are a cross of two breeds are not eligible for shows. It is the pure breeds that are standardized within the Poultry Standards of Europe or USA that are the competitors.

Spoilt for Choice

There are well over 100 of these breeds, so plenty to choose from! If you already own one of these breeds, equip yourself with a copy of the Standards for that breed (and join the club that supports it – they nearly all have one) and take a long, hard look at them. If they look acceptable, then why not have a go at your next local show?

Did You Know?

The National Poultry Club Show of Great Britain, held annually, attracts 5,000 entries representing over 130 breeds of poultry.

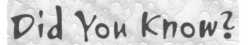

Bring on More Birds

Poultry breeding is an addictive pastime. Whether you choose to use a hen to do the brooding naturally or go for artificial incubation with a machine, there is always a thrill in seeing a chick peck its way out of an egg and the realization of how wonderful nature is. Then there is the anticipation of seeing the chick grow, will she be a good layer, could he be the prizewinning cockerel at one of the big shows or will you be able to market them to fellow poultry keepers for some pocket money to help feed your poultry habit? (*See* pages 232–50 for how to breed your birds.)

Natural Incubation

Most new poultry keepers who keep a cockerel with their hens become breeders by accident. The hen hides up somewhere safe and, much to your surprise, she walks back with several balls of fluff behind her. Or else she is so persistent in her desire to sit on her eggs that her will outweighs yours and you give up and make her safe for her three-week sitting. Either way, you suddenly find you have become a poultry breeder by default and the learning curve steepens.

Artificial Incubation

There are a range of efficient egg incubators on the market that hold around 12 eggs to thousands and have accordingly a wide price range. If you have been running a cockerel with your hens, their eggs will be fertile and you can collect over the course of a week and use the incubator to hatch them.

Of course, then the problem of rearing newborn chicks hatched from artificially incubated eggs is yours – no motherly hen to do it for you – so this is a skill you have to learn before putting the eggs in the machine.

Selling Stock

If you have chosen good-quality birds that are meeting the Standards set down for them, then you should be able to sell their offspring as good examples of their breed. Of course, you can sell crossbred birds too, but there is more value in producing purebreds. It is unlikely you will be able to sell as many cockerels as you raise, even if you sell the birds as pairs – not everyone will want to purchase a cockerel. But there is money to be had from quality purebreds, correctly raised. Maybe not a fortune, but if you do the job properly, you can expect to get some financial reward.

Checklist

No cockerel needed: You can easily keep a few hens as back-yard layers with no cockerel.

Don't believe what you hear: The myths that put people off poultry keeping are just that – myths!

A happy and healthy you: Hens can help keep you and your family healthy through emotional contentment and physical exercise.

An ecological boon: Keeping chickens can be part of a wider green initiative.

Not just eggs to eat: Moving on to meat is a possibility for some keepers.

Birds to be proud of: Showing your poultry is a skilful but inexpensive hobby.

Think before you breed: Breeding is fun, but be sure of what will happen to the new chickens.

Preparing for Your Chickens

Location, Location, Location

It's time to take a long hard look at your garden, back yard or, in some cases, your patio. How much space can you allocate to a chicken house and run? Will they be coming out and free ranging every day, some days or never? Do you have a vegetable area that is only too easily accessible to strong scratchy feet or have you established shrubs that cannot be damaged? These questions will need answering so you can decide which breeds of chickens to keep and so decide the housing.

Size Matters

It really does! With over 100 breeds to choose from in the pure breed poultry world and an increasing number of hybrids, you have to make a choice based not only on your own needs but also on the space available to you in your garden or back yard. It is possible to keep poultry on a patio, but only if you choose a docile breed of bantams (a small variety of poultry), have only a pair or a trio and ensure that the housing provides a sufficiently sized run – not dissimilar to assessing the space needed for guinea pigs or rabbits.

Roam on the Range

You should always provide the maximum space you can for your chickens, not the minimum space they need to survive. Remember that it is natural for all chickens to range, looking for food and scratching for invertebrates. If you are confining them in a run, it needs to be able to allow this behaviour. They also need to dust-bathe to keep their feathers in order and, because they have a pecking order, the ones lower down the order find it very stressful to be jammed next to the top ones.

Check Out the Garden

Look carefully at your garden or back yard for hazards both to the poultry and for anything that they might damage. Ponds are not usually a problem to adult chickens, but are a danger to young birds and chicks. Steep-sided ponds are more dangerous, especially if they lean over to drink.

Fencing

If you plan to let them into the garden for any part of the day, then it will need to be well fenced to keep them in and other animals out. You may need to reinforce the existing fence with at least four-foot high mesh. So choose an area that will work with all this fencing.

A Prime Position

Choosing a site for the poultry house is very important. Consider:

- **Your neighbours:** Do not site the chickens too close to them.

- **Shade and sun:** Chickens do not want full shade, but certainly not full sun either.

- **Drainage:** Does the area get boggy? Make sure the chickens will be on secure dry ground.

- **Convenience:** Do not place them too far away, as you want to be able to reach them easily in all weathers.

Being Neighbourly

Some people are very worried about the idea of their neighbours keeping chickens and feel that they will be noisy, dirty and intrusive. Others will tell you that you cannot keep chickens due to a covenant on the housing estate, planning issues or local area rules and regulations. Mostly, these are just unjustified fears, but still very real to those who have them.

Start As You Mean to Go On

You will need to reassure your neighbours that the chickens will be clean, quiet (that is, not a noisy breed and no cockerel) and will even provide them with some free eggs! It's a really good idea to get off on the right foot with neighbours, who will often find the prospect of a few fresh eggs irresistible and will perhaps be helpful if you cannot get home one night to shut them up.

Check Out Planning and Regulations

In the majority of cases, there is nothing to stop you keeping chickens, but there are a few houses or estates that were built with the deeds stating that chickens or agricultural animals cannot be kept, and, depending on your country or region, some municipalities allow residents to keep poultry and some don't. You should check with your local authorities, zoning and health boards. Also, local planning is another issue. If the house is mobile, then usually chickens are regarded more as pets than agricultural livestock, but if there are complaints from neighbours about noise or vermin, then this will bring in planners to assess how to proceed.

The Agricultural Angle

As chickens are classified as farm animals, different countries have different regulations on registering as poultry keepers, so check this out with your local agricultural department. In Britain, there is a Poultry Register, but this is not statutory unless you keep 50 or more, though you might wish to register anyway, as they send out text alerts of any modifiable diseases. Wherever you live, you must abide by any agricultural regulations relating to poultry which will be highlighted as we go through the book. Briefly these are:

- **Animal welfare**: Any laws relating to animal welfare and specialist laws or advisory notes relating to poultry must be adhered to.

- **Feeding restrictions**: The feeding of food waste is banned in many countries and must be observed.

- **Egg production**: Follow basic hygiene and retail laws, plus any special regulations applying to egg-date stamping under some circumstances – they may not apply to small flocks, but check them out.

Protecting Poultry

So often, when I am advising new poultry keepers, they tell me they don't have foxes. I always reply, 'If you have chickens, you soon will do!' Foxes are not the only predators, but they are one of the most persistent and crafty. There are a number of other furry killers, ranging from the polecat and weasel family (including domestic escaped ferrets), through to badgers who like eggs. In the USA, you can add coyotes, raccoons, fishers and bobcats to that list, and don't forget the threat from the skies with raptors (hawks). Young birds and chicks are particularly susceptible.

Fence Them Out

The first rule is to make sure that predators are fenced out and your poultry is fenced in. Initially, it can be a bit of an investment, but money and time spent doing this will save you money (and emotional upset) in poultry losses. The predator threat does vary from area to area and some poultry keepers may be able to let their birds out in the daytime to free range as long as someone is about, while others will be plagued with hungry killers and need to keep the boundary area well fenced or provide a strong house and large run.

Domestic Violence

Don't let the danger come from within! Pets can present problems for poultry. Usually, the domestic cat is of little danger to a full-sized hen and won't bother to even look at them, but there are a few cats who will take on a very small bantam and a larger number who will try for chicks.

Cowering from Canines

Dogs are more of a problem. A chicken cannot tell if a dog is 'playing' with them or trying to eat them and so are naturally frightened of canines. Keep your dog under close control, on a leash and, where possible, keep them away from the poultry house altogether. If your dog and poultry will be sharing a space such as the back yard or garden, introduce the dog under controlled conditions and take it gradually, small steps at a time. If the dog shows any signs of chasing (even without killing) or attacking chickens, you will have to keep them separate. If the dog is well mannered and ignores the chickens, still remember that poultry do feel threatened by these (to them) wolf-like creatures.

Top Tip

Speak to a professional dog trainer if in any doubt about how to manage your pet's behaviour towards your new chickens.

Children and Chickens

Children usually love chickens and they are hardly 'predators', but this too needs careful managing. Without meaning to do so, smaller children can squeeze or pinch and often shout or squeal, all of which will frighten your poultry. Teach them to handle the birds gently and to be as quiet as they can around them. Older children should already know this, but just reinforce the message from time to time. Never ever allow children to chase chickens to catch them.

Fragile Creatures

Chickens are fragile in their bodies, as are all living things, but a laying hen who is almost always in the process of forming an egg, is particularly so. Crushing her can cause severe and often fatal problems in her oviducts due to internal damage. It is also easy to break legs by accidentally dropping them.

Visitor Alert

Everything that applies to your family applies to visitors. Do not allow other people's kids to chase your chickens and do not let friends' dogs worry your flock. Never feel you cannot correct someone if they mishandle your birds – you are responsible for their safety and have every right to do so. Another reason for good fencing is to stop passing, unaccompanied dogs coming on to your property and hurting your hens. Conversely, be sure your birds cannot get on to other people's gardens where they might not only be annoying to neighbours, but in danger from their pets.

Know Thy Enemy

Get to know the possible predators and dangers in your area and then you can take steps to minimize the threat. Remember:

- **Foxes**: Good fencing and attention to shutting the birds in at dusk every night.

- **Hawks and large raptors**: If your birds are small, choose housing that has a mesh on top as well as on the sides of the run.

- **Polecats**: Check the housing and fencing regularly for small gaps where they can squeeze through.

- **Domestic pets**: Keep yours and any visitors' pets under control. Fence out stray or unaccompanied pets if necessary.

- **Field animals**: These are mostly docile with poultry, but a minority of horses or ponies may be sharp with them – if you own one of these, you will have to confine the poultry.

- **Local threats**: Some areas have predators that are particularly prevalent in their area – ask existing poultry keepers what these are and take steps accordingly.

- **Busy road**: If you live near a busy road, again, fence the birds in so they cannot stray on to the road, putting themselves and road users in danger.

Home Sweet (and Safe) Home

There is no one kind of chicken house, but there are various factors that should be included in them all, such as space, ventilation, perches and nest boxes. Purpose-built chicken houses usually include all of these, but you still need to check, as some are a better (more hen-friendly) design than others.

Minimum Requirements

The minimum requirements of a poultry house are to provide dry shelter, be strong enough to keep out persistent predators even in the dead of night and be able to be cleaned out on a regular basis. It also has to be well ventilated but not drafty and, if the birds are staying in the house for any longer than the night, provide sufficient space for them to move around freely and provide natural light.

House with Run or House and Free Range?

There are three basic types of poultry housing – a house with an integral run, a house and separate but joined-on run or a stand-alone house where the birds are let out into the garden, orchard or paddock for the day.

House with Integral Run

These come in some very imaginative designs, but they are basically a house with a run built into the house. One of the most popular is a house on legs with the area underneath the house wired to create a run. There is a ladder into the house that can be pulled up at night for extra safety.

The advantages are that they can usually be moved easily so that the birds are on fresh ground and they are very safe. The disadvantages are that you are governed by the size of the house as to the size of run, so you need to be very clear about the size of birds and the number that can be housed – do not overcrowd.

House with External-but-joined Run

Here, the house is separate to the run and there will be a pop hole that can be pulled up at night for extra safety. These can be purchased ready-made with run, or you can buy a poultry house and make your own run. The smaller ones are mobile and can be moved around, but for keeping, say, half a dozen full-sized laying hens, it is usually better to make your own run and make it a good size.

There is no specified size for a run, but your birds need to be able to walk up and down and have head room. So, for six medium-sized hens, you really do want a *minimum* of 3 x 2 m (c. 10 x 6½ ft) – that is if they are sometimes coming out to free range – it should be larger if that is their total range area. Ideally, the run should be moveable so that you can rest the ground, or else you may need two runs so that you can rest one and put them in the other – if you are building the run yourself, you can incorporate this into the design.

Stand-alone House

If you are planning to let your chickens free range every day and you have identified a safe area for them to do so, where they cannot be attacked by predators, then you will only need a stand-alone house. There is one caution though. When avian flu was threatening to cause problems, poultry keepers were required to be able to keep their poultry shut in and separate from wild birds. If you have a stand-alone house, just bear in mind that you might need to keep your birds confined on some occasions, so either you need a good-sized house or, more likely, a mobile run to hand.

The area where they range would normally be fenced against predators, although, in some areas, daytime predators are not an issue if there are people around during the day. You do, however, need to have sufficient fencing to keep out passing dogs and keep your poultry from straying on to other people's property.

Top Tip

Don't think, 'what is the minimum space my poultry need', instead think 'what is the maximum space I can give them?'

The Poultry Ark

This is a wedge-shaped, usually wooden, house where one end is solid to provide a house and the other end is the run. A pop hole in the house can be shut at night for extra safety. It was used a lot by commercial smallholders for raising younger birds and is extremely useful, if you are breeding, to house a hen and chicks (though, commercially, chicks are normally reared indoors now). It comes in a range of sizes from really quite small and very mobile, to large and difficult to shift on your own.

Not Good for Big Birds

The thing to remember is that, because the roof goes to a point, the birds do not have head room all the way across the top. If you put a large bird in a small ark, then they can only move along the centre with their head up. This effectively denies them the use of both sides of the run. So, with an ark, get the size right for the birds – unless very large they are not usually suitable for full-time living for large fowl. Many of them do not have perches either, and require the birds to roost on the floor.

Choosing a Chicken House

Remember to think about the following:

- ☑ How much space can I allow for them?
- ☑ What type of chickens shall I keep – large fowl or bantams?
- ☑ How many chickens shall I keep?
- ☑ Are they going to be free range, semi-free range or confined?
- ☑ What type of housing meets these needs and the needs of the chickens?
- ☑ Provide more space than is basically needed.

So What Makes a Shed into a Poultry House?

Chickens are happy in a range of housing, from attractive purpose-built housing to an existing building that has been cleared out and refurbished for them. So what makes a chicken house? First of all, consider the chicken's point of view!

What the Chicken Wants

Put yourself into the mind of a chicken and think about what you would want from a house. Ideally, you would want to go out and scratch and range every day, so you will need space to do that. You will want to feel safe, not only from predators but from your fellow hens, so you will need to be able to get away from them at times, on a perch or behind a box, perhaps placed in the run. You want to be dry inside the house and not to paddle in mud outside the house. You want to be able to flutter up to a perch at night – not too high, but something that gets you off the ground. You want to be able to breathe clean air at all times of the day and night so you don't get lung problems. You want to be able to lay your precious eggs safely and comfortably in a nest situated in a secluded part of the house.

What the Chicken Keeper Wants

The poultry owner wants for his chickens everything they would want, but he or she also wants to be able to collect fresh, clean eggs easily and to be able to clean out the house. If you can stand up in the house, then cleaning out is much easier, but most purpose-built houses have recognized the need to clean and so come with removable sides or roofs, so you can easily gain access from the outside.

The Five Freedoms

There is worldwide agreement on what livestock need in order to be able to live humane lives. These are called 'the five freedoms'. Domestic back-yard chickens should also be treated in line with these standards in all of their care, so consider them right at the beginning, when choosing housing.

1. **Freedom from Hunger and Thirst**: By ready access to fresh water and a balanced diet to maintain full health and vigour.

2. **Freedom from Discomfort**: By providing an appropriate environment, including shelter and a comfortable resting area.

3. **Freedom from Pain, Injury or Disease**: By prevention or rapid diagnosis and treatment.

4. **Freedom to Express Normal Behaviour**: By providing sufficient space, proper facilities and company of the animal's own kind.

5. **Freedom from Fear and Distress**: By ensuring conditions and treatment which avoid mental suffering.

Every Chicken House Must Have ...

Whether you convert a shed or buy a new house, you will need to make sure that the house meets the following needs. Look at them one at a time and you will find that it's not hard to make a house meet the basic requirements of chickens.

Shelter and Security

Look all round the house – can you see any way that a small animal from the weasel family could get in? If so, make it safe. Check for a leaking roof or a wet floor; will the house keep out the weather? If wood, has the wood been treated to resist weather? (*See* page 51.) Will the house remain standing during a strong wind? If not, either look for another house or re-site it.

A Breath of Fresh Air

Chickens are prone to breathing problems if they are kept in stuffy, dirty houses. They will snuggle together for warmth and are good at coping with low temperatures; what they cannot cope with are draughts and lack of air. Ventilation is vital. Hot air rises, so ventilation needs to be situated at the top of the house, above the perching birds. Roosting chickens must not get draughts or, worse still, rain on them. A few small holes are not enough; you need an area that is covered in mesh to keep out predators but allow air in. Don't be afraid to ask poultry house manufacturers how they have provided ventilation and if you are not satisfied with the answer, choose another house.

Perches for Poultry

Poultry prefer to perch, rather than sit on the floor, so make sure you provide enough perch space for the number of birds.

 Perch space: Chickens need a minimum of 20 cm (8 in) of perch width each, for larger fowl, and a bit smaller for bantams. Large fowl are not renowned for their flying skills so there will be a certain amount of juggling and fluttering for them to get comfortable.

Position: Chickens will defecate at night, so do not put perches above food, water or nest boxes. Perches need to be higher than the nest boxes, but they don't need to be really high up – as long as they are off the ground and are all the same height (poultry like to roost as high as they can get, so they will all go for the highest perch).

Nest Boxes

These can be fixed within the house, or on the outside with an entrance from the inside of the house, with lids that can be lifted for easy egg collection or, at their simplest, they can be unfixed boxes positioned in a darker part of the house. All these are suitable and it depends on the house as to which you use. If you are keeping birds in an old large stable, for example, then you might just provide some loose boxes, but a more compact house will have them on the outside to save room.

 Size: Nest boxes need to be at least 30 cm (12 in) square and 15–20 cm (6–8 in) deep, depending on the size of the hens, with a bedding material provided.

 Position: They must always be in the darker area of the house, away from the hustle and bustle of everyday life and out of direct sunlight.

Let There Be Light (and Shade)

The chickens do not want unrelenting direct sunlight, but nor do they want to be spending their days in a dingy dark house and run. Think how you would like a room in your house to be and provide that sort of light. In the summer and the winter, provide shade from sun and shelter from weather in the run.

What Are Poultry Houses Made From?

Wood

The majority of poultry houses are constructed from some kind of wood and therein lies the tale. There is a wealth of difference between a house made of quality wood of human house-building standards and one made of the odds and ends of soft wood. Advantages of wood are that it is versatile (most types of house can be built using wood), comparatively cheap, warm and strong. The disadvantage, as we shall see, is that it harbours pests, including red mites.

 Treating wood: Any new wooden hen house should already be pre-treated with preservative that is not harmful to poultry. The wood then needs treating again yearly – always check that the chemicals you use are chicken-friendly and remember never to put hens straight back into a newly treated house.

Plastic

An increasing number of these are being built because they are easy to clean and don't afford shelter for mites. As with wood, there is plastic and there is plastic. Cheap plastic becomes brittle with frosts and sunlight and won't last. The design needs to allow for the fact that plastic becomes very hot in the summer and needs to provide good ventilation.

Metal

Metal is strong, easy to clean and doesn't hold mites, but becomes very cold in the winter and too hot in the summer. Good designs may be able to overcome these disadvantages.

Brick

An outbuilding or stable can easily be made into a poultry house. It can harbour mites, but is easier to wash down than wood and is strong. Existing buildings are usually bigger than the average poultry house and provide safe space.

Fencing for Fowl

Most chickens, especially the smaller, lighter breeds, are capable of limited flight. Some bantam breeds are quite nimble with their wings and can get into trees. Even with the flight feathers clipped (we show you how to do this later on in the book – it's not painful and won't ruin their beauty), many chickens can quite easily get over a metre-high fence, so it is important to get your fencing right.

Standard Fencing

Choose the best system of fencing that you can afford. It will save you time and trouble later on. Fencing to keep chickens in needs to be a couple of metres (6–7 feet) high, with a folded-over top to prevent climbing predators. You may need to put a wire cover over the top as well. Fencing also needs to be dug into the ground to stop burrowing predators – most wild animals are good at digging!

Let's Go Live – Electric Fencing

Permanent or temporary electric fencing or barriers in a private garden or back yard can play a valuable part in keeping out predators and keeping poultry in. Think about the following:

 Which predators? First, decide which animals are your main threats and then design the fencing to suit. Climbing predators do not like electrified top wires and burrowing

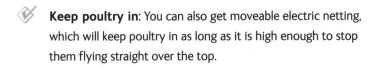

predators will not like a strand just above ground height (not touching the ground, or it will earth and eat up your battery).

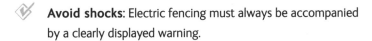

Keep poultry in: You can also get moveable electric netting, which will keep poultry in as long as it is high enough to stop them flying straight over the top.

Power supply: Electric fencing can help in the fight against predators, but it must be constantly live, so you will need a reliable source of power. Solar batteries are becoming increasingly popular in some areas. Get into the habit of checking your fence every morning and use a fence tester to check the current.

Avoid shocks: Electric fencing must always be accompanied by a clearly displayed warning.

Natural Fencing

Tight mature hedges, such as those that surround a garden, might be able to keep poultry in, but cannot be relied upon to keep predators out. This could be a good example of where an internal electric fence could provide the discouragement unwanted visitors need. Remember that predators can climb and drop down as well as burrow.

Top Tip

Unless your fencing is totally predator proof, always shut the hens in their house at night for additional safety.

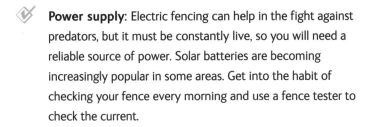

Drinkers and Feeders

Chickens need water and food, which are best given in specially designed containers that have developed over the years to help prevent food waste and to keep the water clean and constantly available. They do come in a variety of designs and some can be hung from the roof of the house, making them less likely to spill.

How Many Do I Need?

The most important thing is to buy sufficient feeders and drinkers so that all your birds have access to them. It is better to buy a couple of medium-sized ones rather than one huge one, as the shyer birds won't eat or drink with the others and may miss out. If there are a couple or more (depending on size of flock), then they will always be able to use them.

Water, Water Everywhere

The reason that most poultry keepers don't use an open bowl is because hens naturally scratch the ground for food, so then bedding from the house and earth from the run quickly get thrown into the bowl and contaminate it.

Avoiding Contamination

Most drinkers are therefore designed to hold the majority of the water under what looks like an upturned bucket which spills into a small circular drinking area. If hung up, they won't be knocked over, but if, placed on the floor, some have legs to stop them getting material in them. Otherwise, stand them on a brick – not so high that the hens cannot reach them, but

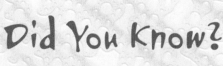

high enough to keep them away from bedding and earth. There is also a bucket with a lip that can be turned on its side and this one I find very effective.

How Much Do They Need?

Chickens drink plenty of water (eggs are 85 per cent water and it needs to come from somewhere), so the drinkers must be big enough to allow clean water at all times. The amount of water drunk varies depending on the weather and the size of the bird – if you allow a minimum of half a pint per bird and put some extra in for waste or evaporation, then you should be about right. If you find the drinker is empty at any time, then increase the amount you give.

Did You Know?

In addition to the main feeders, you will need a container for offering grit to your flock. A small trough will be sufficient.

Thoughtful Feeders

The choice here is a trough with an inward-curling lip which helps to prevent food being pecked out of it and wasted or a custom-made feeder which looks similar to the upside down bucket drinker but has larger holes. This type keeps the feed not in the feeder clean and dry. Outdoor feeders have a sort of hat or umbrella effect over the top to keep out the rain. Finally, there is the auto feeder where the bird pecks at a tag hanging from a string which causes release of the food. This is said to be pest proof as well. (*See also* pages 136–39 for more on feeders and drinkers.)

Positioning

Check out the house and decide where to situate your feeders and drinkers – can you hang them from the roof? If not, position them where they won't get knocked over or walked through. Also, ensure there is enough space not only for all the birds to get their heads round the feeders and drinkers but for shy birds to feed and drink in a more isolated space. Buy two small troughs and put them apart for example.

Inside or Out?

Should you situate your feeders and drinkers inside or outside the house? There was a great deal of debate on this subject at the time of avian flu when it was felt that, due to possible contamination of wild birds, feeders and drinkers should be moved inside and many keepers have continued to do so for this reason. However, birds do need to do something in their run, so it is arguably better to have feeders and drinkers outside. To reduce the risk of

contamination from wild birds, consider the type of feeder carefully – perhaps the auto feeders would be best if using outside in a larger space such as an orchard. If you do keep any water in the house, it may be best to take it out of the house at night if there is any danger of it being spilled – chickens will not eat or drink when it is dark, as they are roosting.

Crested Birds Need Consideration

Some breeds of chickens have a crest of feathers on their heads and these need special care when it comes to feeding and drinking. What you need to avoid is their crests becoming covered in food when they eat and then the food being 'cemented' by water. Choose narrow-lipped feeders and drinkers to avoid this and watch out for any problems.

Checklist

✔ **Size and space:** Think carefully about how much space you can give your chickens – think 'the bigger the better' rather than minimum requirements.

✔ **Optimum position:** Make sure the site of the hen house is dry, convenient and sufficiently shaded from strong sunlight.

✔ **Neighbours:** Reassure your neighbours over any concerns they have and ensure the poultry do not impinge upon them.

✔ **Evaluate the enemy:** Remember that predators can come on wings, can burrow through the ground, can climb and can even swim (if you go on to duck keeping, don't think that the ducks are safe on an island) – be smart and be safe.

✔ **Flight risk:** Be aware that you may need to clip the flight feathers of your birds to stop them flying over the fence (see page 156).

✔ **Shockingly effective:** Electric fencing can be an option, but not in public places and, wherever it is used, it should carry a warning notice.

✔ **Equal rights:** Unless you only have three or four birds, provide a couple of feeders and drinkers to ensure all birds have access to feed and water at all times.

✔ **Another consideration:** If you have to use a bowl for water, then it is better to use a couple of smaller ones rather than one in which the chickens can accidentally get into.

Buying Your Chickens

Choosing Your Chickens

This is a big task, as there are so many breeds and types to consider. Also, you will have to think about what age you want, whether you want a cockerel or not and how many to get. To answer these questions,

think carefully about the space you have available, what you want them for (pets, eggs or meat), how much you want to spend and how experienced you are with livestock – some breeds take more expertise than others.

Cockerel or Not?

For your hens to lay, you do not need to keep a cockerel. If you want to be able to hatch out chicks, then you need a cockerel who will mate with the hen to provide fertilized eggs. Although cockerels are attractive and watching them scratch special morsels for their girls is very amusing, it is best as a beginner to start with hens only, so that you don't have the worry of breeding.

Having a Cockerel Without Breeding

If you do decide to keep a cockerel and you don't want to breed, then you must collect the eggs daily so that they cannot be hatched. This is easier said than done if the birds are free range and wander off to make a nest! They shouldn't do so, through good poultry management standards, but even my bantams sometimes disappear to a safe but hidden spot and come back with a new family.

Hybrid hens do not go broody and therefore will not raise chicks, so they are a good choice if you want a cockerel but do not want to breed. But remember that if you do want to hatch the eggs, you will have to incubate them artificially.

> # Did You Know?
> It is perfectly safe to eat fertilized eggs as long as they have not been sat on by a broody hen for any length of time, which in reality means no more than a couple of days at most.

What Space Do You Have?

By this point, you will have identified how much space you can allocate to your chickens. This will govern how many you keep and which sort you choose. Small spaces require small, docile breeds, while medium to large gardens widen the choice to medium-sized pure breeds, hybrids and ex-battery hens. With a large amount of space, you can think about the larger breeds such as the spectacular Brahma, the iconic Buff Orpington and the now-mainly-exhibited Jersey Giant.

Start Simple

When you first purchase your chickens, begin by having only one sort, such as a hybrid or mixture of same-age hybrids that were raised together, or just one kind of pure breed. It is so easy to end up with many different types and, if you do decide to breed, then it is difficult to provide lots of separate housing for the various breeds. You can soon add to your collection once you have mastered basic poultry care.

What Do You Want from Your Chickens?

Although there are many common characteristics shared by all breeds and types, certain varieties have been bred for quite specific purposes. Human beings have taken the basic model and bred it to produce a higher number of eggs, a larger weight of meat, to be more docile, to be more sparky (in the case of game bantams and fowl), to be able to range economically – that is, to be able to supplement their food by their foraging efforts – or, at the other end of the scale, to reward good nutrition with a very high egg-laying performance.

Bred for Aesthetics

Along the way, other features have been prized, such as extravagantly feathered feet, beautiful head crests and head adornment such as brightly coloured ear lobes or an arrangement of feathers known as a 'muff'.

How Many Eggs Do You Need?

When you first begin with poultry, you cannot ever imagine having too many eggs. If you have six hybrid hens and you feed them properly, you will have six eggs almost every day. That's 42 eggs a week. Omelettes go from being two egg to six egg, you anxiously look up whether too many eggs are bad for you (they're not!), then you look for recipes to preserve eggs (we give some later on in the book), look into selling some eggs (most areas allow farm gate sales, but we look at that later too), or you simply start giving them to (an increasing number of) friends.

Keep it Down

All that is fine, but if you only need a dozen eggs a week, then it might be worth just keeping the number of hens that will provide these, which is two hybrids – they will provide a dozen eggs between them most weeks of the year. I would probably keep three though, just in case something happened to one of them. It means less expense (on chicken feed for instance), less cleaning hassle and less destruction to your garden if you decide to let them out. And you can always buy more hens later on ...

A Reminder

So, when choosing which hens to buy, don't forget to consider:

- How much space you have
- How many eggs you want
- Whether you want children to be able to handle the birds easily
- Whether you want to breed them
- How much experience and time you have
- Whether you want hens that can provide eggs and meat

What Size Chicken?

With over 100 pure breeds and dozens of varieties of commercial birds, from laying hybrids to broilers for the table, to say nothing of crossbred chickens, there is a wide choice of shapes and sizes. The Serema bantam weighs up to 600 g (1¹⁄₃ lb) and is the smallest breed, while at the other end of the scale there is the Jersey Giant, weighing in at around 6 kg (13 lb) for a male. The beautiful Brahma reaches around 5.5 kg (12 lb) and is also a big bird.

Classifying Chickens

With all these breeds, it is no surprise that classifications have been laid down to try to clarify the various characteristics. Most countries use the same categories, which are as self-explanatory as they can be once you understand the technical terms. It is worth taking a moment or two to familiarize yourself with them, as understanding the categories will help in selecting the chickens for your personal situation. The first thing to consider is size classification.

Large Fowl

This is the 'normal' size, but the weight actually ranges from about 1.5 kg (3 lb) to 6 kg (13 lb), so there is a wide variation in size and also build – the more chunky the bird, the more it weighs. So, like any animal, you can get large but light-framed breeds and small but thickset breeds. As we shall see, the lighter breeds are usually the good layers, while the heavier-framed ones are table birds.

Miniatures

Often referred to as bantams, many of the large fowl have a miniature counterpart. They are the same standard as the large fowl but much smaller, usually under 1.5 kg (3 lb) and more

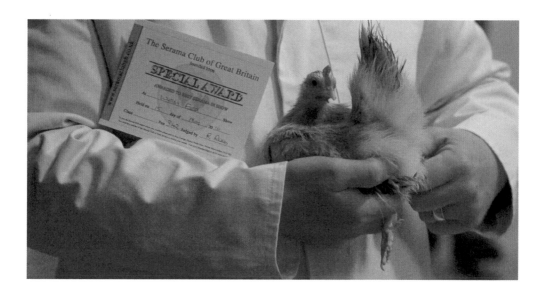

often around 1 kg (2 lb) in weight. It is perfectly all right for all but the most serious show exhibitors to use the term 'bantam' for these, but it is good to know that they have a large-fowl version and are really miniatures of the original breed.

True Bantams

These are, as the name implies, the smaller breeds of chickens, again between 500 g (1 lb) and 1.5 kg (3 lb), but for which there are no large-fowl versions. Some of the oldest breeds are these charming, busy little birds, such as the attractive Japanese Bantam, which are of great antiquity. It is thought that the name 'bantam' comes from the actual city of Bantam in Indonesia, which, along with so many seaports, was responsible for the movement of poultry around the globe from comparatively early times. Boats would take live poultry from the area to provide eggs, with the inevitable result that they reached new seaports and became imported to other countries.

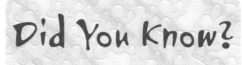

Did You Know?

'Banty' or 'Bantie' is a shortened slang term for a bantam breed.

Further Classifications

Having mastered the basic classifications of large fowl, miniature and bantam, you will learn that chickens are further subdivided by weight and feather type. As previously mentioned, the weight indicates if they are laying (lighter, more active breeds) or table birds (heavier, more docile birds).

Heavy or Light?

'Heavy' breeds are larger and usually more docile than the 'light' breeds. They have the disadvantage of eating more than the light breeds (to support their larger structure) and usually are good broodies (will sit on their eggs and raise chicks). Some of the heavy breeds are good layers, such as the Sussex, but most are more noted for their table characteristics, although the Marans is now really kept for its dark brown eggs.

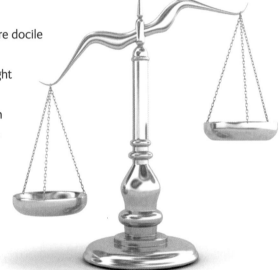

Light Can Be Flighty ...

Some of the most famous layers are found in the light breeds, which are often Mediterranean in origin. But they can also be flighty. The White Leghorn is a good example, having been noted as a prolific layer, but also known for its quickness, speed and liveliness. These sorts of birds do not like being confined in a garden coop. But other light breeds, such as the unusually feathered Silkie, are more laid back and, if handled young, make good pets for children (and their entranced adults!). Most light breeds lay white or slightly tinted eggs.

Feather Type

The classifications here are hard feather and soft feather. Most poultry are soft feathered, with only the game-type variety of poultry being hard feathered. Hard feather chickens include Indian Game, Modern Game, Old English Game and all Asian Hard Feather birds such as Asils, Shamos and Malays. Game birds are the ones that were originally bred for cockfighting, though they are now kept for exhibition and some of their genes have been used for developing some modern hybrids. Their plumage can only be described as tight to the body, which gives them a very glossy, colourful appearance. So tight are the feathers that sometimes there are gaps where they don't meet, which means that some of the skin underneath can sometimes be seen.

Rare Chickens

Some chickens are also classified as 'Rare'. These are pure breeds that are low in numbers. If you choose one of these breeds, they could be more expensive and you will also need to think about breeding to keep the numbers up. They require some research and certainly you need to join a breeders' club.

A 'Citron Millefleur' Booted Bantam – a Rare breed.

Top Tip

Visit a local poultry exhibition or agricultural show to look at the various breeds, and chat to the people who own them for advice.

What Age to Buy?

When starting out in chicken keeping, the usual age to purchase is what is known as 'point of lay'. Whether you do this or go for older birds, it is essential to begin with an adult bird rather than chicks, which require specialist care (*see pages 246–50*).

Point of Lay

This is as it says, or should be, a bird on the point of laying her first eggs. It is actually more difficult to predict or categorize than it sounds. Hybrids will lay at around 18–20 weeks, even in the winter, but many pure breeds will be a few weeks older before they mature to lay and, even then, may not begin to lay until early spring, as most are more seasonally orientated. Most reputable sellers should be able to honestly predict a date when the bird will begin to lay, using their past experience, so do ask them. It is very important to feed these POL (as they are usually known) a balanced ration to allow them to come into lay.

Laying Hens

Into this category fall ex-battery hens and breeds that are in their first year or later of laying. The advantages are that you are purchasing a bird who is already producing eggs and the seller will be able to tell you the sort of number that are being laid. The disadvantage is that the older a bird gets, the fewer eggs they will produce, although

high-yielding hybrids, which include ex-batts, will continue to lay a very satisfactory number of eggs for the household (if correctly fed) well beyond that 'magic' first year.

Young Birds

These will be chickens that have their adult feathers and no longer need outside heat, in the form of lamps, to provide warmth. They will be around 10 weeks old. The disadvantage is that you will have to be patient, as it will be several weeks before they lay, but the advantages are that they will be a little cheaper and you can get to know them from an earlier age. You will need to feed them 'grower' pellets, which will be explained in a later chapter.

Day-old Chicks

These are for the more experienced poultry keeper, as they require heat to replace the hen. In effect, your management will have to replace the mother hen. Most chickens purpose-bred for the table come as day-olds, as their life is only 10–12 weeks, so it would normally be uneconomical to buy them at a later age.

Hen and Chicks

Particularly at poultry sales, you will see hens and chicks for sale and many people do make an impulse buy. If you do, luckily the hen will care for the chicks as long as you supply a safe environment, safe water and correct feed (chick crumbs). The big disadvantage is that around 50 per cent of the chicks will grow into cockerels.

Top Tip

Unless they have already been kept together, never mix ages when you buy poultry. The adults will bully the young ones, sometimes dangerously so.

Hybrid Hens

People talk about hybrid hens as though they are some strange species because they are commercial. Yet they have to originate from pure breeds: selected pure breeds where individuals have particular characteristics, such as good laying, have been crossed with other pure breeds over a period of time to produce 'hybrids'. Common origins are the Rhode Island Red, the White Leghorn and the Plymouth Rock to name a few.

What Do They Do?

They do whatever they have been developed to do. Hens that live in batteries were bred especially to lay a lot of eggs in their first season and to live off as little food as possible (providing it was nutritionally balanced). To a commercial egg producer, feed costs are the biggest input after the cost of setting up, so even a small reduction in intake per bird can make all the difference between profit and loss. Hence why, when you handle a hybrid, they are

light-framed, the food largely going towards egg production and not towards maintaining a big frame. Conversely for meat hybrids (usually known as 'broilers'), they are genetically programmed to want to eat and put on weight, with large frames allowing for plenty of meat when slaughtered.

The Leghorn played a significant role in the breeding of hybrids.

History of Hybrids

The Second World War focused attention on home-front food production. In the 1930s and 1940s, small poultry keepers paid infinite attention to detail, developing high-egg-yielding strains of pure breeds such as the still-famous Rhode Island Red and the Leghorn. Individual hens were assessed with the aid of 'trap-nesting', whereby a bird enters a nest to lay and the door closes behind her, meaning that the poultry keeper knew exactly how many eggs each hen laid.

Evolving Processes

As free range, orchard-based hen keeping gave way to more intensive deep-litter systems of indoor housing, so confinement of chickens continued, leading towards the first battery system. At the same time, breeders, large and small, were trying to improve the egg-laying performance of the breeds and found they could do so by crossing breeds and then crossing again. The battery system and the battery hen were hatched.

Other Reasons for Developing Hybrids

As well as the obvious aim of wanting higher egg production, other elements were considered desirable, such as docility, economic feeding, being a non-sitter and, later on, egg colour; so more crosses were made to achieve the desired results.

Laying Hybrids versus Broiler Hybrids

It has to be mentioned that the cockerels of *laying* hybrids do not fatten, due to the light frame, and therefore hybrids are 'sexed' at a day old by the big hatcheries who breed them and the males destroyed. The *broiler* hybrids of both sexes are raised for the table. Obviously, if you have a cockerel running with your hybrids, they will lay fertile eggs, which you can incubate, but they are non-sitters themselves.

Choice of Hybrids

These days, there is a huge choice of hybrid hens in a range of colours. The Black Rock was developed in Scotland especially for the then growing free-range and back-yard market, a life that suits it very well, but there are now many others with a wealth of names such as Blue Bell, Miss Pepperpot, Warrens, Speckeldy and lots more. They all have a proven egg-laying performance, so the choice is yours for egg colour and looks.

Pros and Cons of Keeping Hybrids

For

- With the correct nutrition and management, they will reliably lay around 300 eggs a year.
- Usually docile and predictable to handle.
- Easy to find to buy and are usually vaccinated.
- Hardly ever go broody, so you never lose egg production.

- Come into lay all year round when POL.
- Available in a range of colours and many are now as attractive as most pure breeds.
- Excellent for new poultry keepers, as straightforward to manage.

Against

- Later on in life, can have health problems connected with the prolific egg laying.
- Must be provided with correct nutrition to maintain their egg laying.
- Hardly ever go broody, so you cannot breed unless you incubate the eggs artificially.
- Purely kept for egg laying, there is no value in preserving them for the future as with pure breeds.
- Some types of hybrids can be a bit sharp within the flock – watch out for pecking or bullying.

So What is a 'Utility' Breed?

The 'utility' breeds were very popular during the war. These were *pure* breeds – not hybrids – that were carefully selected within their breed to produce both eggs and meat, so that the hens could be layers (and then go in the stew pot) and the males could be fattened for eating. So, while they were pure breeds rather than crossbreeds like hybrids, they were selectively bred for similar purposes.

Recent Trends

There is much interest in reviving this type of pure breed these days, but the carefully bred 'utility' lines of such breeds as Rhode Island Reds and Light Sussex have been largely lost, despite retaining the pure-breed exhibition birds. There are just a few poultry keepers left who are keeping and developing these types of birds (including HRH Prince Charles) and they may still make a big comeback.

Buying an Ex-battery Hen

Commercial poultry farmers only keep their hens for their first year of laying, which is the one that produces the highest number of eggs. Yet the birds will continue laying a high number of eggs, by most poultry keepers' standards, for many years. It is possible to obtain these birds either from a battery hen rescue organization or even direct from the farm. As well as batteries, some deep-litter houses and also big free-range units may also have birds available, as they too only keep them for one season.

Battery Hen Basics

If you haven't read the previous section on hybrid hens, then take time to look through it, as hens kept as battery hens are of a hybrid type. They are bred to lay high numbers of eggs to a prescribed nutrition requirement – that is, lots of eggs with correct feeding.

In Good Health

Battery hens will all have been vaccinated in their commercial unit and will usually be currently egg laying if not in moult (see page 224). Although they will be physically very challenged because being caged means they cannot exercise their bodies as nature intended, their internal and external health should be good, that is, not coughing or sneezing and no internal or external parasites.

Appearances Can Be Changed

Battery hens can be alarming on first sight, as, even when not in moult, they can be rather featherless and appear weak – their leg muscles will have been weakened from being unable to do anything but stand in their cages. Because their bodies have become fragile, they need to be handled very gently; they can be damaged very easily. Their combs are normally pale and limp. With care, these birds will soon re-feather, their combs will become smaller and redder and gradually they will learn to range and scratch.

It's a Big Change for the Chickens

If you really think about the lifestyle they have lived, it makes planning their new management much easier. They have been confined, so it will take time for their muscles to strengthen. They have been in a warm shed, so they need to be protected from cold, wet and draughts.

Feed: Help the Transition

They have also had very specific nutrition, so choose a laying pellet ration. It might be possible to buy a bag of their current ration and gradually change them over. Alternatively, many feed manufacturers now make an ex-battery ration, which is designed for the bird to make that crossover from battery to back yard and support their body while they do so.

Getting Confident Chickens

As well as their physical limitations, they will have had emotional deprivation – in other words, they have not been able to directly interact with other birds and have been oblivious to the outside world and usually to natural day and night. They therefore have few social skills and will, to your eyes, squabble, but this is in fact them sorting out their pecking order.

Support them through this by providing several water and feed containers to make sure they all get plenty to eat and drink, and also by providing places where they can get away from each other in their new pen – perhaps a straw bale that they can get on and hide behind. Do not overcrowd them. Give them plenty to do: greens hanging up in the house, pecking blocks (compressed blocks of wheat and vitamins) and logs with bark – both as low perches and to peck at.

The Big Wide World

Mostly being hybrids, ex-battery hens are docile and friendly, and because of their busy environment, not usually frightened by humans. You must be very careful with dogs, as they will be scared of them and also unable to get away from them. Keep your dog on a lead and do not allow them to sniff at the pens with the hens inside. Remember, however friendly your dog, to the hens it is a wolf with bad intentions. It takes time for both sides to learn how to behave with the other and you are the person who must ensure this happens satisfactorily. Do not let children handle your new ex-batts until they are much stronger and then under supervision. Never let anyone chase them.

Venturing Outside

To begin with, simply leave the pop door open into the run and give them incentives, such as food, to encourage them to come out. Even if you plan on having them ranging in a bigger area, do not let them out of the run for several weeks until they are fitter and know where they live. You may have to gently help them back into the house at night, as they will not have experienced 'going to bed' before. Put perches at a low level, as they won't be fit enough to fly up – you can gradually move them upwards. They may spend the first few nights on the floor, so make sure it is dry and warm.

Top Tip

You might like to put a vitamin drink in your ex-batts' water for a few weeks.

Choosing a Pure Breed

Pure breeds have their origins over hundreds and even thousands of years. But some now-recognized pure breeds were developed from Victorian times onwards for many purposes – some just for their pure beauty. At the same time, increased mobility and trade led to different breeds finding their way all round the world. The enthusiasm for Standardizing breeds in the twentieth century continued the trend. All this has combined to mean that there is a huge range of pure breeds, each offering something slightly different, so there is almost certainly the breed for your needs.

How this Book Presents the Breeds

Although breeds of poultry do possess multiple characteristics, such as being a good layer, forager and reasonable table bird all in one, over the next section of this book, for ease of use, the breeds are grouped together by their main characteristic.

Each breed is given its classification – that is, whether it is a large fowl, true bantam or miniature; a heavy or light breed; and a hard feather or soft feather. Also provided are the feather colours available – often there are many, so we have given the most notable ones. If a breed comes in more than one colour, these tend to be Standardized and so are capitalized. Egg colour is given too. There are well over 100 chicken breeds, so the breeds here have been selected for availability and proven use, but do check out the websites at the end of this book for a full list.

The Speckled Sussex breed.

The High-performance Egg Layers

Most new poultry keepers will be looking to get a reliable number of eggs for their household, so we will start by looking at these breeds. Many of the top egg-producing birds can be traced to the Mediterranean breeds of chickens such as Leghorns, which are light breeds and not good for meat. But in the mid-twentieth century, the 'dual purpose' breeds were also popular – they still have a very good egg-laying performance but are also suitable for the table.

Any breed identified as 'dual purpose' will not be quite like the supermarket bird, which is a broiler hybrid but, if fed correctly, will carry enough meat to be a good table bird – so you would have good egg layers in the females and could raise the males as table birds. Also, these days, all of these pure breeds are also kept for exhibition purposes, so you can expect beautiful plumage as well.

Did You Know?

Although smaller breeds usually lay smaller eggs, it is not always the case, and some very large breeds lay quite a modest-sized egg.

Leghorn

Size and weight:
Large fowl; light; there
is a bantam version.

**Feather type
and colour:**
Soft feather; variety of
colours, with White being
the most well known for
egg production, followed
by Brown.

Egg colour:
White.

Originating from Leghorn in Italy but introduced into America around the 1870s. Played
a big part in developing the high-yielding hybrids. Most noticeable is the big comb which
can flop over. This can be subject to frostbite in cold weather. They are a very active
breed indeed, sometimes thought of as 'flighty', and are known for being excitable, so
keep them in a calm environment and treat them calmly. They can fly quite well, so
make sure they are securely enclosed. They do not sit and hatch off eggs, but will lay
200-plus eggs a year, making them one of the best, if not the calmest, pure breed layers.

Ancona

Size and weight:
Large fowl; light; there is a
bantam version.

Feather type and colour:
Soft feather; mottled black
and white.

Egg colour:
White to cream.

A breed of Mediterranean origin, named after Ancona, Italy, where it was developed. It bears a resemblance to the Leghorns in looks and temperament. It is flighty both in nature and in reality, as it can fly quite well, and so will need careful fencing. They react badly to being handled, as they are rather excitable, so not good in a family situation. They need a lot of space to forage and need to be kept busy. They are good layers of around 180 or so eggs, but they do not go broody and, as with the Leghorns, have a very light frame ,which does not make them good for the table.

Hamburgh

Size and weight:
Large fowl; light; there
is a bantam version.

**Feather type
and colour:**
Soft feather; known
for their beautiful
pencilled feathers
(*see* below).

Egg colour:
White.

The breed's origins are unclear, but are thought to be in northern Europe. Despite this, it does seem to have a number of Mediterranean-breed characteristics, being flighty, needing to forage (do not try to confine them in a small area) and not being good broodies. They are respectable layers of around 150 eggs a year, but are not ideal for beginners. They have prominent white ear lobes (though smaller ones than the Minorca, *see* right), but it is the plumage that distinguishes this breed, with the Silver Spangled (pictured) and Silver Pencilled, plus the Gold Spangled and Gold Pencilled, being some of the most striking colour arrangements you will see in poultry.

Minorca

Size and weight:
Large fowl; light;
there is a bantam
version.

**Feather type
and colour:**
Soft feather; Black
(pictured), White
or Blue.

Egg colour:
White.

Another breed of Mediterranean origin, where it was well established before coming to the USA and Britain. It has interesting head gear, with the option of both the single and rosecomb (a low, broad, solid comb covered with blunt beaded points, which goes to a tip) in the large fowl. The female's comb falls to one side, but should not obstruct their vision. It also has huge white almond-shaped ear lobes which make it very distinctive. Frostbite of the comb can be a problem in cold weather. Unfortunately, the concentration on the head features for the show bench came at the expense of the egg-laying characteristics. Even so, it still lays around 180–200 eggs a year, which is pretty good. In common with other Mediterranean breeds, it needs plenty of space to range, is flighty and is not a good sitter.

Sussex

Size and weight:
Large fowl; heavy; there is a
bantam version.

Feather type and colour:
Soft feather; 'Light Sussex'
(white with distinctive black
feathers around the neck and
on the tail) is the best known,
but there is a range of
colours (*see* below).

Egg colour: Tinted.

Also of note: Dual purpose.

A traditional British breed originating, as its name suggests, from Sussex. The oldest
type was the spectacular speckled, and it is thought that the Brahma, Cochin and
Dorking were used to create the Light Sussex, which is the one normally associated
with being a 'dual purpose' bird. Other colours include the rarer Brown, the Buff, Red,
Silver and White.

It is a straightforward bird to keep, being docile, calm and gentle, while also being
active and alert. They will lay around 180 eggs a season, which is below the number they
laid when they were kept commercially during and before the war, but is still a respectable
number. They are also good broodies and will raise their own chicks. Try to buy from a
breeder who can give you some figures on how many eggs their birds produce.

Orpington

Size and weight:
Large fowl; heavy; there is a bantam version.

Feather type and colour:
Soft feather; Black (pictured), White, Buff or Blue.

Egg colour: Brown.

The famous British poultry keeper William Cook introduced the Black Orpington in 1886, the White in 1889 and the Buff in 1894. The Blue came in the 1920s with a rare Jubilee Orpington, now hardly seen. Cook meant these to be dual-purpose birds. However, as spectacularly large, feathery birds, exhibitors wanted to improve on this and so crossed them with Cochins and Langshans, losing much of the commercial performance along the way. So, though truly beautiful birds to look at, they are no longer the dual-purpose breed they once were. Some strains should still be capable of 150 eggs a year, although they are often disappointingly small in relation to the bird's size.

A Good Pet

However, the Orpington is an incredibly loveable, docile bird, making it a great family pet. It is a large bird, so will need a suitably sized house and a run that has height as well as space, though it does love to forage. Because it is so gentle, it is easily bullied by other poultry, so do watch out for that. Because it is so feathery, especially around the vent (where it defecates) you also need to watch out for fly strike, where flies lay eggs and maggots hatch and burrow into the feathers and, worst of all, flesh. It is a good sitter and mother and, although it tramples with its large feet, it is not as scratchy in the garden as some lighter, more active breeds.

Croad Langshan

Size and weight:
Large fowl; heavy; there is a bantam version.

Feather type and colour:
Soft feather; Blue, Black, Buff or White.

Egg colour: Brown.

Also of note: Dual purpose.

The origins of this breed lie in Asia (the Langshan area of China), but how they came to Britain was via the importer, a Major Croad. Confusingly, there is also a Modern Langshan, which is different in shape, with much longer legs, and bears the classification of 'Rare'. In addition, there is the German Langshan, which was bred by the Germans at the beginning of the twentieth century, which does not have the feathered legs of the Croad Langshan.

All-rounder

As far as egg performance goes, they lay around 150 eggs a year from a good laying strain and make tasty table birds. As Croads have feathered legs, they must not be kept in wet conditions (nor should any fowl, but it is especially important for the feathered-leg breeds). They are hardy, placid and enjoy foraging, but can also be kept confined – although they are large birds and need a suitable area. They are friendly and make good family birds. They may not be that easy to find in some areas.

Australorp

Size and weight:
Large fowl; heavy;
there is a bantam
version.

**Feather type
and colour:**
Soft feather; Blue or
Black (pictured).

Egg colour:
Tinted to brown.

Also of note:
Dual purpose.

The name 'Australorp' is a combination of 'Australian' and 'Orpington', so-named
because it was developed by Australian breeders using the Black Orpington – crossed
with Langshan, Minorca and Leghorn. It gained fame in the 1920s with a claim that
a hen laid 364 eggs in 365 days, a truly remarkable record. It made its way into
most countries, including the USA and Britain, with a formidable reputation as a
dual-purpose breed. These days, it produces around 200 eggs a season and is a very
good, white-fleshed table bird. It is also a good beginners' bird, as it is docile, friendly
and, if pen-tamed, good to handle. It is best kept in a large run (remember they are
large birds), but also loves to forage in the garden or orchard. It also makes a great
mother and is a good sitter.

Rhode Island Red

Size and weight:
Large fowl; heavy; there is a bantam version.

Feather type and colour:
Soft feather; rich red-black.

Egg colour:
Light brown to mid-brown.

Also of note:
Dual purpose.

Originating in New England, USA, in the nineteenth century, this breed was developed from Asiatic black-red fowls, Leghorns and probably farmyard fowls of the area. One of the foundation sires of the breed was a black-breasted red Malay cock which was imported from England. The birds were selected for their egg production and size of egg. A straightforward bird to keep, they like to forage and are active, but happy to be domesticated. They are not the best of sitters, but will lay around 180–220 eggs a year. However, try to buy a proven laying strain.

Plymouth Rock

Size and weight:
Large fowl; heavy; there
is a bantam version.

**Feather type
and colour:**
Soft feather; Barred
(black and white) and
Buff are best known,
but also Columbian,
White or Black.

Egg colour: Tinted.

Also of note:
Originally a dual-
purpose bird, but has
been bred more now
for beauty than for
the table.

Developed in Massachusetts, USA, Plymouth Rocks were first exhibited in that country
in 1869 and arrived in Britain a few years later. It is a great back-yard bird due to its
calm, docile temperament, with the breed appearing to be naturally quite friendly and
approachable. It is a good layer, racking up around 160 eggs a year, and is another
breed that will live happily in a large run but is also more than happy to free range.

New Hampshire Red

Size and weight:
Large fowl; heavy; there is a
bantam version.

**Feather type
and colour:**
Soft feather; red.

Egg colour:
Tinted to brown.

Also of note:
Dual purpose; classified as
'Rare', but on the increase as
a back-yard bird.

This breed was developed, as its name suggests, by the farmers of New Hampshire, USA,
but interestingly they used selection from the Rhode Island Red, with no outside breed
introductions, to do so. It was finally Standardized in 1935 in America and enjoyed
something of a rebirth from the 1980s onwards both there and in Britain as a back-yard
bird. It is a good layer of around 150 eggs a season, but that is sadly well down on its
original trials, back in the Thirties, of over 300 eggs in a year. What it does particularly
well, and especially in its bantam version, is to be a good garden or back-yard bird. It's a
friendly, docile breed, the archetypal 'little red hen', and is an excellent mother. It can live
confined to a large run, but is equally happy to free range and forage.

Wyandotte

Size and weight:
Large fowl; heavy; there is a bantam version.

Feather type and colour:
Soft feather; huge range of colours, including Golden Pencilled (pictured).

Egg colour:
Tinted.

Also of note:
Dual purpose.

The very attractive silver-laced variety of Wyandotte was the first to be Standardized in 1883 in the USA and entered Britain at much the same time. It is a large bird, quite upright in appearance and very alert, but docile and calm. It can live in a good-sized run or is happy to forage further afield. It produces 150–200-plus eggs, with the Silver Laced being the best-laying strain. Although they are happy to sit and hatch their eggs, but are not excessively broody, and are devoted mothers.

Vorwerk

Size and weight:
Large fowl; light;
there is a bantam
version.

**Feather type
and colour:**
Soft feather; buff
with black points.

Egg colour:
Cream to tinted.

Also of note:
Dual purpose; Rare.

This is of German origin and was created by Oskar Vorwerk in Hamburg. He wanted to breed a middle-weight, dual-purpose chicken that was economical to keep and lively but not flighty. He also wanted to achieve a breed where males could be reared in comparative compatibility. The Vorwerk today has all these characteristics, making it a great back-yard bird. It lays around 150 eggs a year and is a medium-sized, plump table bird. Sadly, it is rather rare still, but well worth keeping, to increase the numbers and for its laying performance and temperament. Beware – they are quite good at flying, so need good fencing.

The Coloured-egg Layers

The beauty of keeping your own poultry is that you can decide to keep a breed purely for the colour of its eggs rather than the number of them. On the whole, the breeds that produce outstandingly coloured eggs do not lay all that well, but it is often worth the wait. If you keep a pen of hybrids or good pure-breed layers and a few birds that lay coloured eggs, you will achieve eggs every day for cooking and also a supply of comment-provoking coloured eggs for hard boiling or table decoration. The birds below lay spectacularly-coloured eggs!

Does Egg Colour Matter?

Egg shell colour does not affect the nutrition or taste of the egg (and neither does the breed of the chicken, incidentally) – a brown egg is no more healthy than a white egg, so it is really down to aesthetics. The high-yielding Mediterranean chicken breeds mainly produce white

eggs but, with the demand for light brown eggs in the Fifties and Sixties, other laying types, such as the Rhode Island Red, came to the fore. More recently, hens that lay dark brown and speckeldy eggs, such as Marans and Welsummers, became popular, although they do not lay in high numbers. Blue and green eggs are laid by the rather modest layer, the Araucana.

Barnvelder

Size and weight:
Large fowl; heavy; there is a
bantam version.

Feather type and colour:
Soft feather; Black, Double Laced
(pictured), Partridge or Silver.

Egg colour: Brown.

Also of note: Dual purpose.

As the name suggests, this bird originated
in Holland and, along with many other
breeders in the early twentieth century, the aim was to increase the number of eggs from
local farmyard and back-yard fowl. As well as the colourful and reasonable number of eggs
laid, it is a brilliantly colourful bird in appearance, with the Double Laced version being
especially striking. They can lay around 150–170 eggs a year, but the darker-brown-egg
layers tend to be less productive and the egg colour does fade a little as the hens get older.

Back-yarder

The other bonus of this breed is that it is a great back-yard bird, being robust, placid
and friendly – watch out for bullying from other breeds. They will live in good-spaced
confinement, but love to forage as well – watch out for becoming lazy and overweight
if confined. Also watch out for very cold weather, which they do not tolerate as well as
some breeds. Some strains are better sitters than others.

Marans

Size and weight:
Large fowl; heavy; there is a bantam version.

Feather type and colour:
Soft feather; usually Cuckoo (pictured, each feather banded with a light and dark colour similar to a cuckoo).

Egg colour: Really dark brown.

Also of note: Dual purpose.

It often comes as a surprise to its enthusiasts who are keeping it for eggs that this is a heavy table bird which also lays attractive eggs. It is an attractive chicken to look at, but the temperament can be uncertain – some fully grown Marans males can be challenging and, if their behaviour becomes too difficult, then have no hesitation in putting them on the table and look for a pleasanter bird. If you keep them fully occupied, they are less likely to behave badly and they love to forage, so they are ideal orchard or free-range birds. It is difficult to predict the number of eggs, although it should be 120-plus.

Pick Well

There are two main problems with buying Marans. Firstly, not all strains within a breed lay the really dark brown eggs, so you need to buy from a reputable breeder and be prepared to pay extra to obtain that type of bird. You can always breed from them, so it is worth getting the basic birds right. Secondly, some unscrupulous or simply ignorant people will sell the hybrid Speckeldy as a Marans. Although the Speckeldy lays a pleasant brown egg and a lot of them, they are not in the same league as the dark chocolate egg of the true Marans. It is often hard for the beginner to tell the difference, but a Marans, being a table bird, will be heavy-framed, whilst the Speckeldy, being a laying hybrid, will be light, with a sharpish breast bone.

Welsummer

Size and weight:
Large fowl; light; there
is a bantam version.

**Feather type and
colour:** Soft feather;
red-brown to black is
the standard colour.
Silver Duckwing is
also possible.

Egg colour:
Brown to dark brown.

Holland is the original home of this breed, where they were named after the village of Welsum. It is believed that there are many well-known breeds such as the Cochin, Rhode Island Red, Barnvelder, Leghorn and Wyandotte in its make-up. It was Standardized in Britain in 1930. Its egg colour is the most prized aspect of this breed, being described as a deep brown with a reddish tinge, almost a rich, dark terracotta. It's not a bad layer either, with an average of 140–160 eggs a year. Although it is a light breed, the males will make a meal as well. It is a happy, hardy bird that enjoys foraging and is calm and friendly, making it a good family bird. Unfortunately, it is not a good sitter, so to breed from your Welsummers, you will need to buy an incubator or use another breed of chicken.

Araucana

Size and weight:
Large fowl; light;
there is a bantam
version.

**Feather type
and colour:**
Soft feather;
Lavender is
arguably the most
well known from
the range of
colours available
(White pictured).

Egg colour:
Blue to green.

Chile is the origin of this unusual breed, with its even more unusual eggs. First recorded in the sixteenth century, the breed was kept by the indigenous Indians of the Arauca province. It took until the 1930s to get the Lavender established in Scotland, rapidly spreading into the rest of Britain. It is not a prolific layer, but the eggs are bound to be a talking point. It is also an attractive breed itself, with the face covered with 'muffling' and with 'ear muffs'. Despite its unusual characteristics, it is a hardy, adaptable bird that can be kept penned or allowed to forage, and it's a good broody. Watch out for the 'Rumpless Araucana' variety, which is even more remarkable.

Time for the Table

With pure breeds, many of the good layers can also be used for the table, which is particularly useful in the case of cockerels. This final breed in our main list was developed for meat production, though its supporters argue that it is dual purpose due to its egg production record.

Jersey Giant

Size and weight: Large fowl; heavy; there is a bantam version.

Feather type and colour: Soft feather; Black, Blue or White.

Egg colour: Tinted to brown.

Also of note: Rare.

This is the largest and heaviest of the pure breeds. Unlike today's meat breeds, it is slow maturing and bred to range on the farm. This makes it ideal for home production but not suitable for commercial use, as it simply costs too much to feed and house to maturity, which takes up to six months. Compare that with the ten weeks a modern broiler takes.

Increased Needs, Increased Risks

The Jersey Giant is a lovely, docile, friendly breed to keep, but it is large, so housing needs to be sufficient to comfortably accommodate it. It will also require more food and, of course, a balanced ration. It is not an overenthusiastic forager. If kept correctly, it can lay up to 150 eggs a year. It loves to be broody and be a mother but, due to its size, there is a real risk of breaking the eggs or even accidentally harming the chicks, but the better your management (for example, do not disturb her when sitting), the less likely it is that this will happen.

Auto-sexing Breeds

These were quite popular before the Second World War, as an 'auto-sexing' breed's chicks can be sexed by their colour or markings at a day old. The advantage was that the breeder could either destroy the male birds, saving the cost of feeding until their sex could be determined, or raise guaranteed table birds. With the improvements in commercial sexing methods, these breeds fell out of favour, although it is still useful for back-yard breeders to be able to sex easily at a day old.

Examples of Auto-sexing Breeds

The following breeds are all classified as Rare and have bantam counterparts.

Legbar
Size and weight: Large fowl; light.
Egg colour: White or cream, but in the Cream Legbar, it is blue, green or olive.

Having descended from Leghorns, these birds do not make good table birds, so the auto-sexing at a day old is useful if you can bear to humanely destroy the males at this age.

Rhodebar
Size and weight: Large fowl; heavy. **Egg colour:** Brown.

Welbar
Size and weight: Large fowl; light. **Egg colour:** Brown.

Wybar
Size and weight: Large fowl; heavy. **Egg colour:** Tinted

The Welbar breed.

Fancy Fowls and Beautiful Bantams

Even though the breeds discussed up to now are very attractive in appearance and have personality and presence, there are other breeds that have the 'wow' factor, with unusual plumage, head crests and leg feathers, vivid colours and a really star–struck attitude. Although many of these exotic-looking birds still produce a reasonable number of eggs and some have other practical attributes, such as being really excellent mothers, they are mainly kept for their sheer beauty. Here is a selection of such chicken marvels.

A Silver Dutch Bantam cockerel.

Silkie

Size and weight: Large fowl; light; there is a bantam version.

Feather type and colour: Soft feather; White, Black, Blue, Partridge (pictured on next page) or Gold.

Egg colour: Tinted to cream.

The Silkie is a unique breed of poultry for its feathers, which give it such a distinctive appearance. It is actually one of the oldest breeds of poultry and is thought to have come from China, although some people also speculate India or Japan. The earliest account comes from Marco Polo, who wrote of a 'chicken that had fur'. Of course it does have feathers, but very unusual ones that are like hair, giving it the appearance of being covered in fluff. Not only does it have the unusual plumage, but it also has a head crest like a powder puff and, ideally, blue ear lobes.

Deceptive

When people see it for the first time, they tend to think it is very rare or expensive, but actually it is neither, although you would pay more for a show specimen. Though designated as a large fowl, it is small in size, weighing no more than 1.8 kg (4 lb) for a male, while the bantam has a maximum weight of 600 g (1⅓ lb). The Silkie is an absolute delight to keep, a great talking point and makes a lovely pet for children (but handle gently). They are different to other breeds, so do keep them separately. Obviously, they do not want to get wet or muddy, so keep in dry conditions. Watch out for crests getting wet or clogged with food in feeders and drinkers. They are not keen on ranging, so are happy to be kept in well-spaced confinement, as long as they are busy. Note, they cannot fly.

Source Silkies With Care!

Unfortunately, many so-called Silkies that are in fact simply crossbreeds are sold to beginners, so do take some time to familiarize yourself with good examples of the breed before buying. They are exceptionally prone to scaly leg, which is not acceptable, so be sure to examine for this when you buy.

Good Broodies

But they have one more notable characteristic – they are the ideal broody hen, as they live for sitting and raising chicks. This makes them very useful for raising eggs of less motherly breeds. Despite this, in between persistent broodiness, they lay relatively well, perhaps around 100 smallish eggs a year.

Game On!

The birds in the hard-feathered group are what are known as 'game' birds – they were largely the birds that were used in cock fighting. Of course, this is no longer legal and now these birds are happily used for exhibition purposes or, depending on the breed, as pets. They have also contributed to many of the other purebred poultry breeds and extensively to hybrid breeds, especially in the case of the Cornish and Indian Game, who are excellent table birds in their own right.

Bad Rep

The fighting reputation of many of the breeds is a bit of a handicap, as they can be lovely birds to keep – the Old English Game Bantam is

A Wheaten Carlisle Old English Game.

truly delightful and very good with children. Other breeds may take more skilled management and it is never a good idea to allow game birds to mix with other poultry, just to be on the safe side. Unfortunately, some breeds of game birds attract two-legged thieves, so you need to keep them away from the public eye and well protected.

Some Examples

Hard-feather fowl are one of the most spectacularly coloured groups and all come in a wide range of Standardized colours, with names such as Jubilee, Wheaten, Duckwing and Birchen. Here is a brief introduction to the most popular game (hard-feather) varieties:

Indian or Cornish Game

Size and weight:
Large fowl; heavy; there is a bantam version.
Egg colour: Tinted.

Excellent table bird, but poor layer.
Very broad breasted. Not normally
considered a novice's bird. The
pictured bird is the Jubilee colour.

Old English Game Bantam

Size and weight: Miniature.
Egg colour: Tinted.

This is a fairly recent breed and derives
from the large fowl Oxford and Carlisle Old
English Game. It is a delightful bantam,
absolutely packed with personality, but not
as aggressive as its name suggests. They
seem to stay in tight family groups and are
naturally inquisitive and friendly. They also
seem to tame easily. The charmingly small
eggs can number about 100 or more in a
season, but they are superb sitters and
fierce mothers. They are an incredibly
noisy breed, even the females, and must
be busy – they hate being closely
confined. Hardiness, robustness and
courage is bred into them. The pictured
bird is the Spangled.

Modern Game

Size and weight:
Large fowl; there is a
bantam version.
Egg colour: Tinted.

This is a long-legged game
bird which was developed
for exhibition purposes.
Although there are large
fowl, it is the bantams
that are more widely kept
and shown. Similar in
temperament to the OEG
bantams, these are not
such good layers, but are
impressive in looks. Not
really a beginner's bird.
The pictured bird is the
Black Red colour.

True Bantams for Beauty (and Some Eggs)

We've looked at the bantams that are miniatures of the large fowl, but what about the
'true bantams', the ones with no large fowl counterpart? In this category come some of
the most flamboyant chickens in looks, yet with a useful egg-laying record and great
temperaments. They come in far too many colours to list!

Pekin

Egg colour: White or cream.

This is a lovely garden breed, delightful for adults and very good with children if well handled. It comes in a wide range of colours (Black pictured) and is a diminutive, feathery ball with feathered legs. Docile yet active, this breed is not as destructive in the garden as some breeds. They lay relatively well and are good broodies. Because of their feathered legs and lowness to the ground, they should not be kept in muddy conditions.

Sebright

Egg colour:
White or cream.

One of the oldest British varieties, known for its strikingly marked plumage. It comes in two colours, Gold and Silver (pictured), but it is the uniform and attractive lacing that marks out this breed. Poor egg-laying performance. It can fly quite well.

Dutch Bantam

Egg colour: Tinted.

This tiny breed, which is ideal for small gardens, is an impressive layer of around 120–150 small eggs a year. It is an active little bird and loves to forage, which can be a problem, as it is also quite a good flyer. The pictured bird is Gold.

Japanese Bantam

Egg colour: White or cream.

With their short legs, silky loose body feathers and extravagant tails, these little birds are outstandingly beautiful in looks. Unusually, they come in three feather types: plain- or normal-feathered, silkie-feathered and frizzle-feathered. They are poor layers and do need specialist care as they must be kept clean, not overcrowded and kept away from any bullying birds. The pictured bird is a Black Tailed Buff Frizzle.

Serama

Feather colour:
They come in a range of colours, including White, Black (with blue-green sheen), Buff (pictured, Frizzled), Red, Partridge, Wheaten, Mottled, Spangled, Duckwing.

Egg colour:
White to brown.

Imported into the USA in 2001 and the UK in 2004, the Serama can be traced as far back as the 1600s and originated in Malaysia. They are the smallest chicken in the world, with the male weighing up to 600 g (1⅓ lb) and the female up to 500 g (just over 1 lb) – little more than a pound or half kilo of sugar. Their calm, friendly temperament means they respond to handling, but keep it gentle, bearing in mind their size. They need extra special protection from predators, pets and any rough handling and should be housed in a secure house and run – only ever let out into the garden when you are there to supervise. Keep them protected from wet weather and very low temperatures. They are not productive egg layers but that's not the reason why you would keep this diminutive, delightful bird.

Did You Know?

There is a breed known as a Frizzle, which is a large fowl with a bantam counterpart. But other breeds can also feature frizzle feathers. 'Frizzle' refers to feathers that curl towards the bird's head, which gives the bird's plumage a curled look.

How to Find Your Chickens

Having decided what sort of chickens you would like to keep, how do you now go about finding them to buy? What should you look for in a seller and where can you find a breeder?

Buying From a Breeder

In the case of pure breeds, the very best option is to buy direct from the breeder, so you can see the parents and related birds and really get to know the breed you are buying. Be honest with the breeder in your requirements. If you want to breed in any form, even for your own pleasure, then you need to buy good examples of the breed. Tell the breeder if you think you would like to exhibit one day. These birds will be more expensive but worth the investment.

If You Don't Want to Breed

All breeders have less good examples too, which are perfectly adequate for back-yard birds and which, to less-tutored eyes, still look absolutely gorgeous and carry the characteristics (laying, temperament and so on) of the breed. These will be cheaper.

A good breeder will sell you good examples of the breed, with healthy feathers, such as here.

Buying From a Dealer

There are many reputable dealers, but it is often hard for the beginner to know which ones they are. A good dealer will have a tidy yard with birds kept in clean conditions and not overcrowded. The birds will be alert and show signs of good health. He or she will be able to tell you some history of the birds, such as whether they are a laying strain, how they were reared and if there was any show success. Dates of worming and any details of vaccinations should all be available. If you have a problem with your chickens, a good dealer should be available to talk you through it or, in extreme cases, take the birds back and replace them or refund. He or she should not sell you something that is unsuitable for your needs.

Whoever you buy from, the birds should be in good health. Here, you can see a bird with healthy eyes.

What to Avoid

A poor dealer will have lots of overcrowded pens with poor-looking birds and will not seem to know much about the breeds they are selling. It is not acceptable for a dealer to sell you birds with external parasites (look under their wings for telltale signs or, if you are itching after handling the birds, then they are suspect). They should not have scaly leg at any age (less scrupulous dealers will tell buyers that all feathered legged birds get scaly leg – not with correct management, they don't!). They should not be dull-eyed, certainly not coughing or swallowing noticeably (suspect gape worm) and must not be dirty round the vent (back end).

Buying From a Market or Auction

Let the buyer beware! There are some excellent, graded sales where poultry is carded with a grade card prior to sale and the breeders are on hand to guide you through your choices. There are also some good-standard poultry auctions with knowledgeable auctioneers.

Be Wary

But there are some very rough auctions too, where the previous owners are not available for conversation and the birds are elderly, poor quality or even sick. If you do buy from any sale, then you need to quarantine your birds for at least a week away from any other poultry you may have. Never buy separate pens or different breeds and then expect to mix them when you get home, as there will be bullying. It is better for beginners not to buy from anywhere but a specialist graded poultry sale where help is on hand.

Hatcheries and Agents

Hybrids are hatched and reared in very large numbers at a hatchery and then agents will sell the birds. They are usually very reliable with a reputation to consider. Do not be put off if they are not local, as many do deliver in special air-conditioned trucks. Birds are usually vaccinated and in very good condition. You will not be able to view the birds before delivery, but the agent will be able to guide you through the choice of hybrid.

A healthy hen.

Direct From a Farm

Your local poultry farm may sell off its first-season layers – we have discussed the care of these birds and the preparation you would need to make for them. Contact the farmer direct and ask if you are able to purchase a small number.

From a Charity

Many animal charities are seeking homes for poultry. They are often very keen to rehome cockerels, but do not be tempted to have more than one (and remember, you don't need to have *any* for hens to lay and be happy). They may not be too knowledgeable on breeds and many of their chickens may be crossbred and aged. On the other hand, specialist ex-battery hen charities should be able to rehome birds as well as offer with plenty of advice. Beware unscrupulous dealers posing as a charity who are asking for substantial 'donations' for the birds – ask for their charity registration number and to see their accounts.

Where to Look for a Seller

Sometimes, it seems a daunting task to find chickens for sale in your area, but it's not as difficult as it seems.

- **Breed Clubs**: Contact the Poultry Club for your country or region that looks after poultry Standards and they will give you a list of clubs for each breed, then contact the specialist club for the breed of chicken you are looking for and ask for a list of breeders.

- **Magazines**: Look in specialist poultry and smallholding magazines for advertisements from breeders and dealers, plus adverts from hatchery agents.

- **Local adverts**: Check out the notice board of your local agricultural merchants or feed store.

- **Internet**: Look on the internet for details of breeders and sellers near to you, as well as charities.

Transport and Settling In

You've made your choice and are now a poultry keeper. How do you get your birds home and how do you settle them into their new surroundings? The watch words here will be preparation in advance, and calm, quiet handling on their arrival.

Getting the Poultry Home

There are rules and regulations surrounding the transport of poultry, but mostly they are basic common sense.

 Security: The boxes or containers need to be strong, with plenty of air. Do not place them in the boot of the car and ensure they are secured within the transport so they do not tip up as the vehicle goes round corners.

 Comfort: The birds must not be transported in too hot or too cold circumstances. Unless you are driving a long distance, you do not need to worry about watering the birds but, if the weather is hot and the journey a long one, you may need to stop and try to give water. Be careful not to let the birds escape. Remember, it is a stressful experience for any chicken, so you need to make it as easy as possible.

Regulations

Basic official regulations for transporting poultry include the following provisos:

- **No one shall transport animals, or cause them to be transported, in a way likely to cause them injury or undue suffering.**
- **Journey times must be kept to a minimum.**
- **The poultry must be fit to travel.**
- **Those handling the poultry must be competent.**
- **The poultry must not be injured in loading or unloading from the boxes.**
- **The containers must provide sufficient space and be secure.**

Introducing Chickens to Their New Home

The house and run must be ready for your chickens before they arrive. Carefully put them into their new surroundings, either by gently lifting them out and placing them in the house or run or by opening the lids of the boxes within the run so they can make their own way out. Be quiet and gentle, do not allow noisy visitors, children or pets nearby while you are doing this. Do not let them out of the run for the first few days or weeks – depending on how well-fenced your garden or paddock is, the chickens should be kept confined to their run.

Food and Drink

Water and feed should be placed where they cannot knock it over as they explore the house and run. Keep an eye on them just in case they do and you can soon replace it. After they have had their initial look round, you could get some mixed corn or wheat and give them a few handfuls to scratch at and talk to them while you do this. You want them to associate you with pleasant experiences.

And So to Bed

If you introduce them to their new home in the dark, then place them in the house and keep the pop hole shut, but be ready to open it first thing in the morning. Let them make their way out of the house – don't hurry them. If putting them in the run in the light, then help them into the house at dusk if they don't go in by themselves – give them the chance to make their own way in first. They will soon work it out and do it on their own.

The First Few Days

Your chickens will need peace and quiet for the first few days, to get used to their new house. Introduce the rest of the family quietly and gradually.

Food and Drink

Try to use the same food as the previous owners and change them over gradually to your bagged ration. Give them a 'scratch' feed in the afternoon and talk to them. As I've mentioned, with ex-battery hens, it might be a good idea to add a vitamin supplement in the water for the first few days.

Keep an Eye

When you think they have settled in, which will take at least a week and maybe longer, you can let them out for an hour before bed into the garden if you want to, but keep an eye on them. Make sure that they are all eating and drinking and watch out for any signs of disease, such as sneezing, dullness or scratching. You will also need to be on the lookout for possible bullying.

Egg-spectations!

When should you expect your new chickens to lay? It really does depend on their age and breed but, for hybrid 'point of lays', between 18 and 24 weeks old – they at least need to get settled in their new environment first before they will start laying. They will start with small eggs which will get larger and they will lay all over the place to start with – you need to encourage them into nest boxes by situating them in the darker area of the house and placing a 'pot egg' (a special dummy egg) – or a golf ball would do – in each nest.

Pure Breeds and Laying Hens

Pure breeds are more seasonal, so you might have to wait until the days lengthen. For birds already in lay, there will be a period of a few days while they recover from the stress of being moved (though they might lay one egg that was already being produced in their body before they moved). The sooner they settle in, find the food and water and get to know you, the sooner you'll get your first fresh eggs.

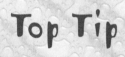

Top Tip

Perhaps phone the breeder to let them know how your chooks are getting on – it's good to keep in contact.

To Name or Not to Name?

It is usually not a great idea to name anything you are going to eat, especially for children. Adults tend to cover up their emotions by naming table birds with rather prophetic names such as 'Drumstick' or 'Curry'. But laying hens are another matter. You can name them for your family, your friends or wait and see what their personality is. Here are some real-life poultry keepers' comments.

A Chicken of Courage

'We named our chicken "Emily" because she was attacked by dogs and almost died; she was so brave I named her after Emily Pankhurst. And "Wee, Wee" because, when his foster mother had had enough of him, he used to run round in circles going "wee, wee, wee"'.

An Eclectic Mix

'"Bumble", because we had just noticed the first bee of the year in the garden and, in the way that chickens do, Bumble raced into the picture and plucked it out of the air and ate it; "Spike", because she was sprayed accidentally with the purple dye we use when clipping the sheep's hooves and it made her look like a punk; and "Ping", because it means "small" in Chinese dialect.'

Don't Let Children Name Chickens!

' "Ben Hen", "Tasty Geezer", "La Roux" – we no longer let the children name them! We have a trio called "Fred", "Ginger" and "Ava".' 'My teenagers named ours "Rock Chick", "Biker Chick" and "Punk Chick".' 'My children named our chickens "Pinky", because my daughter loves pink and didn't know what else to call her, plus "Nelly", because she was the size of an elephant. We also have a "Roxy"... Scott, my son, named that one!'

A Literary Slant

'We have one called "Perdita", meaning "the lost one", because she was my friend's dad's bantam, and went missing from his allotment; eventually, she turned up again, but by then he no longer kept chickens so we had her. She must be at least 10 years old and still laying! We also had a "Henrietta" for obvious reasons. "Big Bum Betty" who really is the chicken with the biggest backside. "Henny Penny" from "the story with Foxy Loxy in it". "Ginger Ninja", because she is ginger and fiesty. "Speedy Gonzales", because she is the fastest chicken on two legs ... she belonged to a school I was working at and no one could catch her, 'til Oliver (youngest son) rugby tackled her under a blackberry bush. "Scooby" and "Dizzy" already had names when they were given to us, and finally "Hedwig" is the white cockerel who resembles the owl in Harry Potter. Reading back through that, I am thinking maybe we are slightly mad.'

Pun-tastic

It is often fun to name chickens according to a chosen theme and create puns based on characters or people associated with that theme. For instance, freelance writer, photographer and podcaster Emma Cooper is a keen gardener and lives in Oxfordshire with husband Pete and several chickens. Names are based on a Star Wars theme and have included 'Princess Layer', 'Chewbucka', 'Cluck Skywalker' and 'Hen Solo'.

Some Traditional Names

- Older lady names, such as Matilda or Martha
- Flower names, such as Primrose, Clover or Lily
- Names to do with colour, such as Snow White or Rose Red
- Doris
- Harriet
- Nancy
- Charlotte
- Daisy
- Laura
- And of course ... Henrietta

Checklist

- **Think before you buy**: Decide your MAIN reason for wanting chickens – eggs, pets, beauty or meat. How much space and time do you have available?

- **Choose a breed carefully**: Read this section carefully to check out the types and breeds available. Do some extra research by visiting poultry shows.

- **Buy wisely**: Try to buy your pure breed birds from a specialist breeder who can show you other related stock or, if buying from a market or auction, choose a reputable one with experts on hand to help your decision.

- **No need for cockerels**: You do not need a cockerel for hens to lay.

- **Safe with hybrids**: Hybrids are guaranteed eggs layers if fed correctly

- **Ex-battery hens**: These require tender loving care initially, but still have many years of egg laying left.

- **Utility breeds**: The old-fashioned utility pure breeds were the ultimate dual-purpose birds.

- **Colour is personal**: Eggs provide the same nutrition whatever their colour.

- **Not just a pretty bird**: Fancy bantams can be useful egg layers as well as being beautiful and suitable for smaller gardens.

- **To transport poultry**: Use strong, safe boxes with plenty of air that don't shift about in transit.

- **On arrival home**: Settle in your new chickens with the emphasis on peace and quiet and do not let them roam free until they are accustomed to their new house.

- **Choose your names for your chickens carefully**: Check out our suggestions.

Feeding Your Chickens

How a Chicken's Gut Works

A chicken's digestive system is so different to that of ourselves and our other livestock that it is worth taking a few minutes to understand the basics of how it works. Then the feeding (and later on, management) advice will make more sense.

Begin at the Beginning – the Beak

Hens have no teeth and, we think, few taste buds. They have to rely on size and texture to select their feed. They also have to swallow food whole, helped by a little saliva and their tongues. You will often see them throwing back their heads to help get the food down into the oesophagus. Beaks are made of keratin, the same tough protein that makes our nails and horses' hooves. They grow all the time, as they clearly need to renew themselves because of wear and tear.

The Crop

The oesophagus leads to the 'crop', which you can see, when you stand in front of the bird, as a bulge (larger when full) at the front of the mid-neck. If you run your fingers down the neck of the bird gently, it is easy to feel. It is basically a container for food. Whilst in the crop, the food takes in water before continuing on down the oesophagus towards the stomach area. This is divided into two bits, the 'proventriculus' (or 'glandular stomach') and the 'gizzard'.

Did You Know?

The saying 'as rare as hens' teeth' derives from the fact that hens do not have teeth and means that something is very scarce indeed, if not unavailable.

The digestive system of a chicken

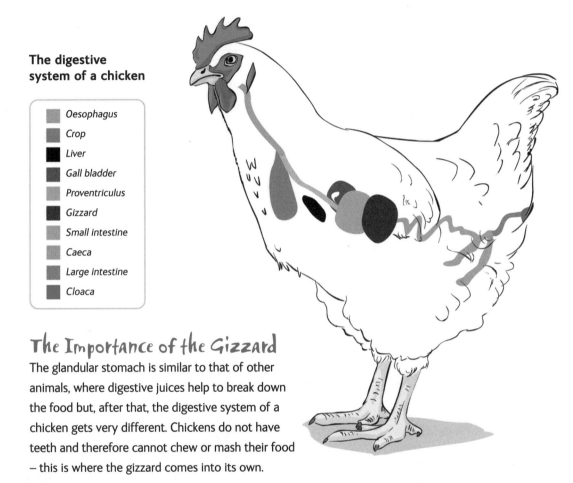

	Oesophagus
	Crop
	Liver
	Gall bladder
	Proventriculus
	Gizzard
	Small intestine
	Caeca
	Large intestine
	Cloaca

The Importance of the Gizzard

The glandular stomach is similar to that of other animals, where digestive juices help to break down the food but, after that, the digestive system of a chicken gets very different. Chickens do not have teeth and therefore cannot chew or mash their food – this is where the gizzard comes into its own.

The Role of Grit

It is a strong, muscular stomach that acts as a grinder and it can only operate if the chicken has taken in grit. Whilst free ranging on grass, the bird will do this quite naturally but, when confined, supplies of grit become harder to come by, so it is essential for the poultry keeper to supply grit. These tiny grit particles act as minute millstones, finely grinding the food into a paste so it can move on down the gut.

Taking in Goodness in the Small Intestine

As the food passes into the small intestine, the paste is broken down by enzymes into its component parts and then the absorption of nutrients into the bloodstream takes place, allowing the chicken to use these to keep in good health and produce eggs. It is here that bile is added to help digest fat. The pancreas and the liver also help to ensure the chicken gets all the goodness it needs.

And Passing out Waste

The 'caeca' is a passage that breaks down waste food before it then travels to the large intestine, where more water is taken on board and finally the waste is passed out of what is commonly known as the 'vent', but which is really called the 'cloaca'. The cloaca has many tasks, being the only opening in the bird's body – as well as eliminating food waste, it also delivers eggs and takes in semen. Chickens do not urinate separately and any uric acid comes out as the white matter you will see on top of the brown faeces.

High Input Means High Output

For their relatively small stature, chickens have a high food intake and a fast metabolism. The higher yielding the chicken (for example, the faster growing, the heavier a table bird, the more productive an egg layer), the more food they have to take in to keep up with their body's demands.

The result of this is that a high amount of food waste is also expelled – an average bird produces up to 150 g (⅓ lb) of faeces per day, 75 per cent of which is water. They defecate anything

between 20 and 50 times a day. They also excrete a frothy brown excrement a couple of times a day, which is a result of the fermentation in the caeca – it's easy to mistake this for diarrhoea but, as long as it is infrequent and the chicken is healthy in other ways, it's perfectly natural.

Are Chickens Vegetarian?

Not by choice! They are omnivores like us, and can eat vegetable and animal protein. These days, strict food regulations mean that it is usually illegal to feed any form of meat directly to chickens, so no carcasses, waste food or even the contents of a sandwich should be given. This is to prevent cross-contamination of disease and should be strictly adhered to.

No Need to Feed Meat

Old-fashioned books will talk about feeding meat to raise the protein ration but, with today's balanced feeds, you will not need to do this. Of course, chickens can get some 'animal protein' from foraging for insects such as worms, slugs and snails and you'll see them suddenly jump up and catch a passing fly with real relish.

Top Tip

From beak to vent can take as little as four hours for young birds and about eight hours for adult chickens, so digestion is an ongoing, fairly speedy process. To keep this marvellous machine in full working order, it is vital to feed digestible, well-balanced feed on a regular basis.

What's on the Menu?

Feeding chickens has become much simpler with the development of bagged feed. The improvement in nutrition (along with genetics) is one of the reasons that laying performance and meat production have risen. The green poultry keeper can benefit from this, knowing that, by using a balanced food ration, their birds will be healthy and productive. We call these 'compound' feeds, and single feedstuffs, such as wheat, are called 'straights'.

Feeding Naturally

These days, bagged compound feeds do not contain antibiotics as routine, nor are they allowed to include any animal protein since the link with BSE was proven. Chick-rearing feeds may include anti-coccidiostats (ACS; *see* page 212), which are not always a bad thing but, if you have clean rearing conditions and you are not rearing on a large scale, then you may wish to choose a ration that does not include these and you can do so.

The other two considerations are whether any of the ingredients are genetically modified or whether you go for an organic ration. Both of these types of compound feed are readily available.

The GMO Puzzle

Chickens require a balance of carbohydrate, fat and protein and the proportions will differ according to whether you are keeping high layers, table birds, pets or exhibition birds. Protein is usually supplied in the form of soya, which is a crop associated with GMO (genetically modified organism) seed.

Therefore, it is not possible for many compound-feed manufacturers to guarantee that their feed is GMO-free. But it is still possible to choose a GMO-free ration, as there are many millers who produce this type of feed and it will be stated on the bags. It will cost extra (but not excessively so), as it is harder to source guaranteed non-GMO ingredients.

The Organic Option

If you want to take it a step further, look for a certified organic ration. This will have been produced using organically grown ingredients, although, in some rations, a percentage of non-organically grown ingredients is allowed. You can check on the label. Expect to pay more for an organic compound feed, as the basic ingredients cost more to buy – organic wheat, for example, is often twice the price of non-organic wheat, as the yields are lower.

Top Tip

All organic feeds should carry a certification symbol or guarantee from a recognized organic certification body. Look for this on the bag or label. You can contact the organization direct if you want to know more about the composition of the feed.

Different Feeds for Different Birds

As there are several reasons for keeping poultry, so there are several different types of feed that will be appropriate for different ages and uses of chickens. All will be nutritionally balanced for those individual needs, including containing vitamins and minerals.

Let Them Lay

It is incredible that some breeds or types of chickens can produce the complex and demanding object that is an egg almost every day. The layer needs plenty of nutritional help to do this. If she does not get it, then she simply is incapable of laying an egg. The ration will

include extra vitamins and minerals to keep her healthy, plus calcium for the eggshell formation, as well as a balance of protein, carbohydrate and fat.

Start Them Young

From hatching until about 6–8 weeks, chicks are fed on a starter ration more commonly known as 'chick crumbs'. These are higher in protein than the other rations, as the chicks will be growing very rapidly, and they are smaller in size so the chick can peck them up easily. They may or may not contain anti-coccidiostats (ACS; *see page 212*) – the choice is yours, so check the label carefully.

Growing On

As chicks grow older, at about 6–8 weeks, they will gradually be transferred to a grower ration, which contains a little less protein than the starter ration but still more than the adult feed. It will also continue with the appropriate vitamins and minerals for fast-growing youngsters.

Poultry Breeder

This is a higher protein ration for breeding birds and has increased levels of vitamins, minerals and amino acids to give optimum fertility and egg quality, which will help to increase hatchability.

Ornamental Poultry

This feed normally has extra levels of oils to improve the plumage. The protein will be lower than for layers or breeders.

Pellets or Mash?

What Are Pellets and Mash?

- **Pellets are very small, compacted rolls of the ration.**
- **Mash is a powder that may or may not be mixed with water.**

The majority of poultry keepers feed pellets, as they are convenient, do not spoil as easily and are usually cleared up well by the chickens. But the argument for mash is that fed dry, not wet, it is more difficult for the birds to peck up and so keeps them occupied for longer. Not many people feed wet mash these days and pellets are nearer to the size and type of food selected by chickens naturally. If you do feed wet mash, then you have to throw away waste mash at the end of every day and mix a new one on the following day, or it will sour and cause digestive upsets.

Different feeds: layers' pellets (top), wheat (bottom left) and mixed corn (bottom right).

Grains and Seeds

Although the compound pellets or mash will contain everything your birds need, most poultry keepers will also want to feed some kind of corn or grain feed, as the poultry really enjoy it; plus, as it takes longer to digest, it makes a great late-afternoon feed.

Wheat

Whole, cleaned wheat is much enjoyed by chickens. Be careful if you buy what is known as 'tail corn' direct from the farmer, as it will contain weed seeds and, more seriously, may not have been properly dried and could be musty. Tail corn is what is left when wheat is put through sieves to filter out weed seeds and broken and small grains of wheat – so, really, it is rejected matter.

Maize

Feed this as cracked maize or maize grits, not as whole maize. Chickens absolutely love this and it's a great winter feed, as it is known for its warming characteristics.

Barley

Not seen so much now in mixed corn, as it is not really enjoyed by chickens, though they will eat it reluctantly. It is high in carbohydrate.

Oats

Another grain that provides warmth in winter, as it's digested slowly, providing warming carbohydrates.

Peas and Lentils

Peas must be fed as split peas and are high in protein, though not always really liked by chickens. Lentils can also be used.

Sunflower Seeds

These are expensive, so use sparingly, but they are much enjoyed and, as they contain a high level of oils, will make plumage gleam.

Mixed Corn

You can mix this yourself using the grains above as 'straights' – if you do this, use mainly wheat and vary the other ingredients according to the time of year. This does not replace the compound ration. Otherwise, feed manufacturers sell bags of mixed corn, which are nutritionally balanced and now many are developing a type of 'super mixed corn' under varying names, which not only have a good balance of straights but also have added oils and even vitamins and minerals. These are much enjoyed by chickens, but they do tend to pick out what they like best first, such as the maize.

Can I Mix My Own Feeds?

The short answer to this is: with difficulty, if you want them as nutritionally balanced as bagged compound feeds. You would need to buy in a variety of straights and you would then need to know the exact protein, carbohydrate and fat content of each one. In addition, you would need to buy in vitamins, minerals and amino acids to add to the mixture and even then you would probably find that the birds eagerly ate some of the grains in your mixture and left the others, thus unbalancing your ration. With the vast improvement in bagged feeds in the last few decades and the choice of non-GM and organic added to the mix, it is better for you and your birds to ensure they eat a nutritionally balanced mix that will not vary.

Oyster shell

Crushed oyster shell is a source of calcium, which is needed by laying hens to enable them to form shells on their eggs. But if you are feeding a compound feed, the necessary calcium will already be built into that ration, so there is no need to supply this separately. It is not the same as the grit needed for the gizzard.

Grit

As I have mentioned, the gizzard must have grit to grind the food, as chickens have no teeth and therefore cannot chew their food. They can get this from free ranging, but confined chickens must have access to grit, usually fed in a grit hopper or a container with holes underneath that allows water to drain. Sometimes, grit is sold mixed with oyster shell – your chickens will self-select what they need.

Time for Titbits!

Feeding kitchen waste is illegal in the UK and varies in the rest of the world, so check out your local laws. You must never feed meat waste of any kind. Garden waste such as fresh green vegetables are very acceptable and can be hung in the hen house to provide something for the hens to peck at. The official line is that bread is not suitable for poultry but, like many things, fed in moderation it will do no harm. Never ever feed any food that has gone mouldy, as it is as bad for chickens as it is for us.

Green Food

Greens are not actually as essential as the main rations and grains, but chickens love to scratch and to forage, so if you let them free range at least part of the day, or even at weekends when you are around, they will benefit enormously. It is difficult to keep grass or vegetation in the run, although moving to fresh ground and having more than one run will help. You can hang up cabbages, Brussels sprouts stalks or other greens for them to peck at and it is a really good idea to grow trays of grass to put in on a regular basis. Do not feed grass clippings, as they may cause compaction of the crop due to the length.

What's on the Label?

All bagged feed, both straights and compound feeds, will have a detailed label, which is interesting to take time to read – it will include:

- **Protein:** Typically around 14–15 per cent for layers, and higher for growers and chicks.
- **Fibre:** Higher for adult birds.
- **Oils:** Another source of energy.
- **Ash:** Represents the mineral content of the food.
- **Vitamins:** Such as A, D and E.
- **Medication:** Usually drugs such as ACS (anti-cocciodiostats, *see* page 212.).
- **Certification:** If it is organic or GM-free.
- **The weight of the product**
- **The actual ingredients:** Such as wheat, soya, barley, sunflower seeds, limestone and yolk pigment (personal preference whether you like to have this).
- **Best-before date:** If it is exceeded, it usually does not mean the product cannot be fed, but that the vitamins and minerals can no longer be guaranteed. But it is best not to accept a product that is past its 'best-before' date.

Feeding and Watering

There are many ways for poultry keepers to give feed and water to their chickens, with a variety of containers on offer. The key factor is that chickens must have access to both feed and water at all times of the daylight hours.

Water

It is essential that the chicken has access to clean water – without it, she will be unable to function, to digest food and, of course, to lay. Even a comparatively short time (such as half a day or so) without water will start to compromise the chicken's body and this will be seen in her laying performance.

Choosing Containers

Containers should be designed to keep water clean while dispensing fresh supplies. They need to be stood off the ground, otherwise it is so easy for chickens to scratch bedding and earth into the container and soil it. If they are overcrowded or the containers are badly situated in the house or run, then the chickens will also defecate in them and knock them over. Be observant and, if this happens, move the containers until you find a safe place for them. Stand them on bricks so that they are off the floor and less likely to get clogged with bedding or soil.

 Domed or tube: The traditional round, domed (or tubular) drinker releases the water into a round trough at the bottom. Its domed top makes it less easy to perch on top and thus defecate into the water. Some 'mushroom' shaped ones with a pointed top have the same effect, as well as making it difficult for chicks to stand in the water. (*See* the image on page 54.)

With legs and handles: Some drinkers have the added bonus of legs, which are helpful for keeping the drinker raised above ground. Often, the legs are adjustable. Wire handles are another plus point, making them easier to carry or providing the option of hanging.

 Bucket drinker (*see right*): These are solid and sturdy buckets made out of galvanized steel, which you fill up as you would a normal bucket and then turn on their side (they have a flat panel across the top, which gives a flat edge), whereupon the water becomes available behind the lip of the panel. Again, the rounded shape prevents perching, so avoids soiling of the water.

 Tripod drinkers (*see left*): These are big plastic or metal containers that stand on long legs and can be filled via the screw top – they are more suitable for big flocks of large fowl.

Water from the Tap

Do not be tempted to use water from a water butt unless it is very fresh, as this can be contaminated by wild birds drinking or defecating. It also becomes sour and stagnant in hot weather.

Feeding

As we have already seen, there is a huge choice of types and sizes of feeders, but it helps to understand the aims of a feeder, which are to supply food ad lib while keeping it clean and dry. In addition, the feeder should prevent waste where possible and allow safe access to the chickens.

Basic Types of Feeders

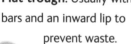

- **Flat trough:** Usually with bars and an inward lip to prevent waste.

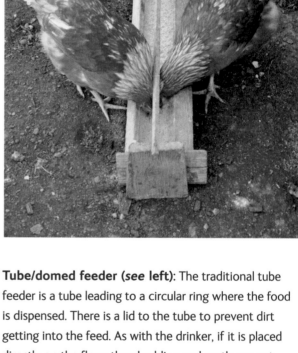

- **Tube/domed feeder (*see* left):** The traditional tube feeder is a tube leading to a circular ring where the food is dispensed. There is a lid to the tube to prevent dirt getting into the feed. As with the drinker, if it is placed directly on the floor, then bedding and earth can get scratched into the food, contaminating it.

- **Gear-wheel feeder:** This is where a tube feeder has strips in the dispenser, which are designed to stop the food being flicked out and wasted.

- **Hat feeder (*see image on page 55*):** This works in the same way as a tube feeder, but the lid extends into a 'hat', so if this feeder is put outside, it will act as an umbrella, keeping feed dry (except in really driving rain).

- **Wall feeder:** As the name suggests, this is attached to the wall, off the floor, and also has bars to stop food wastage. It also features a lip that curves inwards for the same reason.

- **Auto feeder:** One example being the Parkland auto feeder, this is great outside, as the hens peck at a tab hanging down, which releases the feed that is always kept dry and fresh. It can take a while for some chickens to grasp the idea, so watch out for those that are not therefore getting fed.

- **Spiral feeder:** This features a spiral of wire through which the chickens peck at the grain. Often used for pheasants.

How Much Should I Feed?

Chickens should be fed ad lib; they should never be without food. Watch carefully how much they eat and feed accordingly: if they run out, then feed more; if the food becomes stale and sour because there is too much, then feed less.

Top Tip

Feed the compound feed, such as the layers' pellets, in the morning when they are hungry, so they fill up on this and therefore take in all the nutrients they need, and then give a 'scratch' feed of mixed corn in the afternoon, so they have something in their crop before they go to roost.

Store Feed Safely

Any type of bagged feed needs to be stored away from rats, mice and wild birds and in a dry, cool place. Your chickens will not be the only living thing, keen to have the benefit of that nutritionally balanced ration!

Get a Storage Bin

A metal dustbin with a well-fitting lid is the safest bin to choose, with heavy-duty plastic containers running a close second. Really determined rats, and even mice, can eat through plastic, so do check any such bin for holes on a regular basis. Mice not only consume your feed, but urinate rather enthusiastically, which will contaminate it.

Check the Condition

When you tip the bag into the bin, check that the feed smells appetizing, is not dusty or mouldy and that the bag has no telltale vermin holes. Feed merchants will always take back any mouldy or unsatisfactory feed, as long as you report it to them soon after you have bought it.

Checklist

Know the gut: Get a basic understanding of a chicken's digestive system to understand their needs – remember that the gizzard is the grinder and needs grit in order to function.

Drinking: Water must be available during all daylight hours.

Compound feed: Choose the right compound feed – don't forget there is a choice of types, including GM-free and organic. Compound feeds come as pellets or mash – pellets being the most widely used.

Straights: Single ingredients such as wheat are known as 'straights'.

'Scratch feed': Chickens enjoy grains and seeds and these can be fed as a 'scratch' feed.

Check the label: Look at the label for contents, ingredients, weight and best-before date.

Multiples: Provide more than one feeder and more than one drinker, unless you only have fewer than five or six hens.

Cleanliness is key: Choose containers that keep water and feed clean.

Assess your chickens: Watch carefully to judge how much to feed your birds.

A.m./p.m.: Feed compound feeds that are nutritionally balanced in the morning and grain in the late afternoon.

Storage: Use storage bins to discourage rats and mice. Always check food is fresh and fit to feed.

Caring for Your Chickens

A Commitment in Time

How much time does it take to look after your chickens? Well, it's actually not so much the time but the daily commitment that is the main challenge. Although keeping a few chickens may only take a few minutes on a daily basis, it has to be a twice-daily routine, every single day. Then there are weekly and seasonal jobs too.

Take Time to Enjoy Them

Although in a well-designed house and run, your poultry will be quite happy with a few minutes of your time in the morning and evening when you check them and top up water and feed, the more they see of you, the tamer and friendlier they will become. You too will get more out of your new venture the more time and effort you put into it.

Just Stand and Watch

You will learn a lot just by watching the way your poultry behaves. It is important you recognize the pecking order of your flock; who is the dominant hen and which are the shyer ones. The head of the pecking order has to be the strongest, as she will be expected to be the defender. The order will be established right down to the most subordinate bird. If you introduce new birds into a group, then you totally wreck that pecking order and it has to be established again, sometimes with violent and even fatal behaviour.

Boys Will Be Boys

If you have a cockerel, then he will be in charge, calling his girls over to show them a choice morsel of food and then standing back so they can eat it. He will not easily tolerate another male and, if you do have more than one cockerel, they will have to work out who is 'top bird'. Some poultry flocks do this reasonably amicably but, in others, it results in real violence and one of the males has to be removed. It does depend on age and breed.

Getting to Know Your Chickens

You cannot tame chickens just by getting hold of them and handling them; in fact, it will probably terrify them. What you will need to do is to get them to know you by speaking to them, being with them and, most importantly, hand-feeding them special titbits of food. Some individuals will quickly realize that you are good news and come to you, while others will hang back but, the more time you spend with them, the tamer they will become. (Though there are certain breeds that are quite naturally less tame than other breeds).

Top Tip

Time spent observing your birds' behaviour is never wasted.

A Daily Routine

Chickens require at least twice-daily care every single day and, if you are away on holiday or out for the day, then you will need to arrange someone to do it in your place.

First Thing in the Morning

In the summer, this will be when you get up, probably around 7 a.m. but, in the winter, it will be when it is light.

Assess: As you approach the house, check that there are no signs of predators trying to break in or any wind or weather damage.

Release: Let the birds out into their run or on to their fenced free range area. As they come out, check to see that they are lively and alert, which are signs of good health. Any birds that do not come out into the run need checking. They might just be laying or they might be going broody or they could be sick.

Supply: Check the drinker – it will need to be full of fresh water for the day ahead. Check the feed containers and refill, being sure not to put fresh food on old or sour feed. Replace any greens with new ones. Check that there is sufficient grit in the grit containers.

Freshen: Replace with fresh bedding if the bed is damp.

Collect and clean: Collect the eggs and clean out any droppings in the nest boxes.

Final check: Have a final look at the birds to check they are eating, drinking and alert before you leave.

Duties at Dusk

The chickens will go to roost at dusk and that's when the foxes really come out hunting, although some may well be on patrol during daylight hours. If your chickens are on free range, it is vitally important to get the pop hole closed as soon as the birds have gone to bed, to prevent them being dinner.

Shut the door: Shut the birds in their house at dusk when they have gone to roost. Obviously, this will vary from early afternoon in winter to very late evening in summer and will play havoc with your social life!

Head count: Check all birds are in the house.

Risk assessment: Have a good look round the house to check it is predator-proof.

More eggs? Collect any eggs laid during the day.

Remove obstacles: If the food containers are in the run, remove overnight so they do not attract vermin. Empty water containers ready to refill in the morning.

Weekly Care in Addition to Daily

There is usually more time available at the weekends, so that is when you can fit in the extra tasks involved in being a chicken keeper, such as cleaning, checking chooks and just having more time to observe them.

Cleaning Out

Chickens do defecate frequently and so need regular cleaning out. You will need a stiff hand brush, a scraper (a paint scraper will work well), a small shovel and a dustpan for smaller houses, plus a full-sized broom, shovel and perhaps a fork of some kind for a larger setup. You will also want something for the muck, such as an open sack, a garden carrier or a wheelbarrow.

Some hen houses come with a slide-out dropping tray, which makes cleaning simpler.

On a Weekly Basis

☑ **Clear out:** Take out droppings and soiled bedding – pay particular attention to the nest box, as a clean nest box means clean eggs. Scrape perches.

☑ **Compost:** You can compost the poultry manure, but you will need to mix it with other composting waste to make a garden compost.

☑ **Clean drinkers and feeders:** Remove the drinkers and wash them thoroughly – an organic washing-up liquid will do a good job and be quite safe. Thoroughly clean the feeders, but do not put them back wet or food will sour.

Check for invaders: Check round the housing for any signs of mice or rats or any signs of predators scratching at the run or house.

Check the food: Make sure you have enough food for the following week and check that the storage is secure and vermin have not found a way into your storage bins.

Check health: If your birds are reasonably tame and can be caught without fear of injury, it really helps if you can handle each of them to check weight, look for signs of external parasites and look closely at their eyes, legs (for scaly leg), combs (frostbite in winter) and beaks.

Check the hens' environment: Look critically at the run – is it getting muddy and, if so, what can you do to improve it? Hang greens and check that the run provides boxes or straw bales so that birds can be at different heights to allow timid birds some space.

Free range and enjoy: For some families, this may be the time you can let your birds out into your garden, as you are there to deter predators and to watch over them while they are out. It is also time to enjoy your poultry, take time to watch them and to engage with them.

Seasonal Tasks

There are tasks that need doing on a seasonal basis – those that need doing regularly but not weekly, or that are relevant to certain seasons.

Spring Care

 Discourage broodiness: This is when pure breeds think of breeding and, even without a cockerel, hens will start to show signs of going broody. It is very important that eggs are collected twice a day, and even more often if possible at this time, to try to deter them from this. We will cover breeding later on, but watch out for chickens hiding themselves and sitting – if on free range, they might put themselves in danger from predators.

 Keep the run dry: Spring often brings a lot of rain, so watch out for the run becoming waterlogged and, if it does, then look to provide another area. If that is not possible, you will need to put in bark chips, gravel or slabs to give the hens a dry area.

Summer Care

 Keep your chickens cool: Make sure the house does not heat up through being positioned in full sun. Ensure there is shade for the hens – provide branches, straw bales, even a material cover if necessary, so hens can get out of full sun. Provide extra water containers, so the chickens do not go thirsty. Let the birds out of their houses as early as you can, as long as the area is safe from the fox.

 Watch out for the moult: Keep an eye out for laying hens losing feathers and continue to feed nutritionally – perhaps add extra support in the form of soluble vitamins. Some birds will moult in the early autumn.

 Mite prevention: Although lice and mite prevention should be carried out all year, the summer is the peak time for mites, so keep on top of them using mite-control products in the house and anti-mite powder or spray on the chickens and nest boxes.

Top Tip

Use the summer as an opportunity to do any repairs necessary to houses and runs.

Autumn Care

Weatherproof? Time to make sure that poultry houses will stand up to the wind and rain of autumn.

Autumnal fare: Introduce 'warming' afternoon scratch feeds of maize.

Winter Care

Beat the freeze: As with summer, water again becomes an issue. Water containers should be emptied at night and refilled in the morning – check later on in the day as, on very cold days, water can freeze in a matter of an hour or so.

Use your discretion: Chickens are usually reluctant to come out on to bright white snow, so you may have to encourage them out but, if the weather is really snowy, wet or cold, let them stay inside.

In the house: Make sure that they have easy access to food and water. Remove any frosted greens. Ensure bedding is dry, as the chickens' feet may have trampled in snow.

Frostbitten? Large-combed birds suffer from frostbite, but rubbing petroleum jelly on them should help to prevent this.

Hygiene for Health

The two do go hand in hand – to achieve good health, the environment needs to be clean, well ventilated and large enough to allow the chickens to behave naturally. Choice and cleanliness of bedding is very important. It also aids the control of red mite.

Which Litter?

Choice of floor covering is often based on what is most freely available, but some litters are easier to keep clean than others and have composting advantages. Ammonia from soiled bedding will interfere with the bird's respiratory function and will lead to disease. Another cause of respiratory problems could be poor ventilation, birds shut up in a stuffy environment. If you have the two together, your chickens will soon be sick.

Wood Shavings

The preferred litter amongst most poultry keepers, these need to come from white, softwood to be absorbent, and hardwood can cause splinters. They are easy to clean out and the resultant manure will compost, although it will take longer than straw.

Sawdust

This can be very dusty and that is the last thing you want in the confined space of a hen house, but it can make a great dust bath outside in a covered run.

Wood Pellets

These have recently been developed for pet bedding and are relatively expensive, but they are super absorbent, with a pleasant smell. They are larger than

the feed pellets and I have not seen them being pecked up. The disadvantage is that they do roll around when the birds walk about.

Straw

The traditional litter for poultry, but it has some drawbacks – make sure that the straw used is clean, bright and free from mould spores. Straw baled in a damp autumn will be very musty and mouldy. Secondly, it is not as absorbent as shavings and needs regular turning and wet patches removed.

Hay

You are unlikely to use this, if only because of price alone, but it does not make good bedding either because, unless it is very well made (and particularly pricey!), it can harbour mould spores and dust which again is bad for respiratory health.

Purpose-made Beddings

The horse world has leapt forward in this area, using a number of dust-extracted beddings from a variety of sources, such as hemp, many of which are very suitable for poultry, as they are absorbent, dust free and easy to manage. Check out your local feed merchant.

Top Tip

Chickens need to 'dust bathe', so you should provide a tray of sand or, ideally, very dry earth. The tray must be big enough for them to stretch out their wings. They also use their litter inside the house sometimes.

Handle with Care

Chickens are very fragile and must be handled correctly so as to protect their internal organs and bones. The practice of picking a hen up by its legs is completely wrong and can cause severe damage to her oviduct, resulting in death, or dislocate her legs and hips.

Softly, Softly Catches Hen

Handling and catching is so much easier if your hen is not terrified of you, and this can only be achieved by gentle handling from an early age, seeing and interacting with the poultry keeper and a regular routine. Being quiet around the chickens and avoiding sudden movements also helps them to get confidence in you.

Time of Day

If you have to catch a number of birds to move, for example, wait until they have gone to roost, then you can quietly remove them from their perches with the help of an assistant to shut the lids of the boxes. They can then be transferred to their new home and let out first thing in the morning as you would when you get new birds.

Catching a Chicken

Try to avoid having to 'catch' chickens by getting hold of them at dusk as we have said (*see* below), but if you have to catch in the daytime, get the chicken into a confined space and herd her into a corner with your arms outstretched, then make a decisive move.

It is not acceptable to chase chickens round in a large space; you will get out of breath and annoyed and they will get more and more terrified and stressed. Plan where you are going to catch them – perhaps lure them into a small space using corn, or simply don't let them out of the run if you know you want to handle them.

How to Handle

Begin learning how to handle a hen by picking her up when she is roosting. Pick her up with both hands, taking control of her wings, as flapping wings can hurt your face, but perhaps more importantly, the wings run the risk of damage. Bring her towards your body. It is the same if you lift her out of a box, except you can slip your arm under her, as she is contained by the box.

Lifting a Chicken Out of a Box

Slide your hand underneath her body from the front to the back so that her breast rests on your forearm and upward palm and your fingers can gently separate her legs. (These should be between your gently closed first and second and third and fourth fingers.) Do not squeeze tightly, as this may damage her internally and upset her breathing.

Meanwhile, your other hand is resting over her wings but, when she feels safe and balanced, you should be able to remove this hand and use it for examining her whilst she sits comfortably and securely on your arm and hand with her head looking towards you.

Wing Clipping

Some breeds are quite good at flying, certainly enough to get out of their pen or run or to make their way into a neighbour's garden. You can fence them in securely or you can consider clipping their wings.

Is it Cruel?

'Clipping wings' really means to clip off the tips of the main flight feathers on one side. This is not to be confused with 'pinioning', which is mainly practised on wildfowl (some duck and geese varieties) and involves removing a segment of bone as well, and is done when the birds are ducklings or goslings. There are arguments for and against pinioning – it is certainly not necessary for domestic chickens. As for wing clipping, it is up to you, but consider the pros and cons:

For

- It is painless and very simple to do.
- The flight feathers will grow back.
- The chickens don't have to be fenced in over the top as well as the sides.
- They can come out in the garden without danger of fluttering into next door's.

Against

- They will be more vulnerable to predators as they cannot flutter up to get away.
- It will be noticeable and spoil the line of their wings.
- You cannot exhibit a chicken at a show with clipped wings.

How to Clip the Wings

Handle the bird carefully as previously explained. Get someone to hold the bird and then you can spread out the wing and find the flight feathers, which are the two slightly longer feathers at the tip of the wing.

You trim these feathers, staying about 4–6 cm (2 inches) away from where the feathers grow out of the wing – this way, you will avoid where the quills of the feathers have blood in them. Cut about 8 to 12 feathers. This is enough to simply unbalance the bird so that when it tries to take off, it will not be able to do so.

How Often Should They Be Clipped?

Clip the feathers when they grow back again, which is going to be after the moult. Do not cut immature, still-growing feathers, as you may draw blood. Make sure the feathers have finished growing.

These are the main flight feathers of a hybrid hen before being clipped.

After clipping.

Hen Health

It is important to recognize the signs of good health in a hen. Watching the chickens as they come out of their house in the mornings is the best time to spot the signs, although you will need to also handle the hens on a regular basis to check more closely.

Behaviour Says a Lot

As well as the outward signs of health, observe the behaviour as well. Watch and ensure they all want to eat, see who is lower down the pecking order and check for bullying; are any sneezing or opening their beaks wide and 'gaping'? Are there any traces of diarrhoea (not to be confused with the caecal dropping previously mentioned in the section on feeding)?

In the Morning, Look For:

- Alertness
- Bright eyes with no discharge
- Red comb
- Comb and wattles should not be damaged
- Feathers should be tidy and in good condition
- Feathers should not be missing (except in the moult)
- The vent should be clean and not be soiled with droppings
- Legs should be clean and the chicken should not be lame
- There should be no discharge from the nostrils
- The bird should be eagerly looking for food
- On return to the house at night, the crop should be visibly full

Prevention is Always Better than Cure

Understanding the needs of the chicken and providing for them are key to keeping your hens healthy. Many diseases are caused or allowed to develop due to poor housing, lack of ventilation, muddy soiled runs, incorrect nutrition, lack of clean water or overcrowding leading to stress. Although good management won't mean you never get a health problem, it will seriously reduce the likelihood and, in the end, not only save suffering in your small flock but also save you money.

Seasonal Risks

Seasons and weather can catch out even the best poultry keeper, but watch out for extreme weather. Getting ready for the change of season and thinking of the challenges faced by your chickens will help you to prepare for any possible problems.

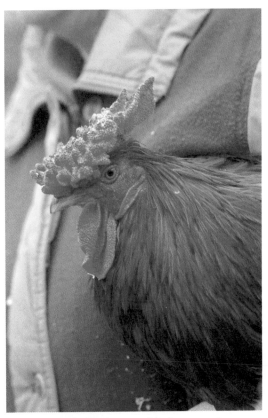

The Derbyshire Redcap has one of the largest combs, making them vulnerable to frostbite.

 Comb frostbite: Chickens with big, fleshy combs can be quite susceptible to this, which manifests as black edges to the comb. This is dead tissue which will drop off. To prevent this, grease the comb with petroleum jelly and ensure the house is draught-proof. Cockerels are particularly susceptible due to their larger combs and the fact that they do not roost with their heads tucked under their wings as hens do.

Heat stress: This shows itself as panting, dullness and the bird being very uncomfortable. Do not situate the run and house in full sunlight, watch out for first thing in the morning when you might be in bed and not notice, do not shut chickens in a small space in hot weather and pay particular attention to very feathery birds such as Cochins.

Should Your Chickens be Vaccinated?

Most hybrids, when you get them, will routinely be vaccinated for Mareks disease. Mareks is a herpes virus that normally shows itself as paralysis of the legs and wings or as neck twisting. It is spread by inhalation and so spreads rapidly through large flocks, hence the vaccination of hybrids, who are usually reared in large groups.

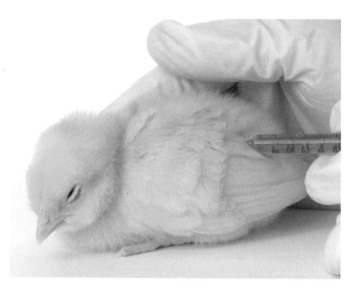

Did You Know?

The life expectancy of a full-sized chicken (large fowl) is six to ten years, while a bantam can expect eight to ten years. Hybrids, due to their high-performance egg laying ability, tend to suffer more from problems with egg laying, which may lower their life expectancy, although you would still expect four to eight years.

There are pros and cons for vaccination, with some breeders disagreeing with it, as they feel it weakens resistance. However, those with large flocks tend to advocate it, feeling it is protecting their poultry.

If you have purchased already-vaccinated hybrids, then you do not need to worry about Mareks. The majority of garden and back-yard poultry keepers who keep a small number of birds in an area that has no history of infection will also not have a problem. For anyone else, it is best to consult your local poultry vet for advice.

First Aid Kit

As with humans, it is best to have some things already on hand. These should include:

- **Petroleum Jelly**
- **Liquid paraffin**
- **Obstetric lubricant**
- **Sharp scissors, large and small**
- **Nail clippers**
- **Cotton wool and buds**
- **Surgical type disposable gloves**
- **Antiseptic spray, which is sometimes coloured blue or purple**
- **Stockholm tar**
- **Aloe vera gel**
- **Louse powder**
- **Mite spray**

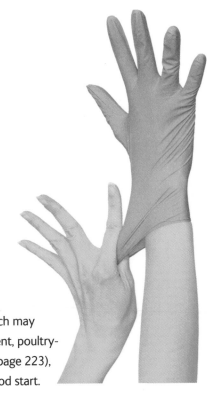

Additional Items

You'll collect some extra items as and when needed, which may include worming liquid, vitamin supplements, eye ointment, poultry-orientated disinfectant and perhaps poultry saddles (see page 223), but if you have the basic kit, that will get you off to a good start.

Veterinary Help

Do not attempt to treat anything but the most straightforward ailments yourself. When you first start keeping chickens, you will need professional advice until you are more experienced (and for some diseases, you'll always need skilled help). It may cost to have a vet consultation, but you will find you can do more and more on your own as you progress, so look upon it almost as a 'setup' cost. For your chickens' welfare, do not hesitate to consult a vet if you think you need to do so. Don't delay in the hope the chicken will just get better on their own – they nearly always get worse if they are not treated promptly, and the earlier they are treated, the more chance of survival.

Breathe Easy

Respiratory diseases can be due to a virus or bacteria, but poor ventilation and overcrowding, leading to a stuffy house, will predispose birds to infections or cause breathing problems of their own. Plenty of fresh air and no draughts will go a long way to prevent respiratory diseases.

Mycoplasmosis (Once Known as Roup)

An infection that presents as a sneezing chicken with discharge from the nostrils. The breathing sounds laboured, probably rattly, the sinuses are swollen and there is a foam-like discharge in the eye. It is highly contagious and can be carried on your clothes, so if you visit a show or other poultry, change your clothes before handling your birds. You will need veterinary help to diagnose and prescribe suitable antibiotics. The house will need washing out with a disinfectant to kill the bacteria, and affected birds should be kept warm but not in stuffy conditions, away from other birds.

Aspergillosis

This is caused by inhaling fungal spores from mouldy or decaying plant matter, such as damp hay or decaying mouldy straw. It is usually caused by inadequate ventilation and by dirty conditions, often coupled with too many chickens for the space. Chickens in severe respiratory distress will need to be culled. Improve ventilation, clean out the house and reduce the stocking rate (that is, reduce the number of birds to the amount of space).

Infectious Bronchitis

Caused by a virus, this also reduces egg laying in older birds. Chicks can be vaccinated against it. Treatment includes consulting your vet for antibiotics and keeping birds warm, well fed and with good ventilation.

Egg Laying Health

New poultry keepers often worry about things such as their birds becoming egg-bound (failing to lay the egg they have made) but, in reality, most chickens lay frequently and successfully as a result of good management.

How to Help the Laying Hen

Correct nutrition, including the inclusion of calcium in the ration, a well-designed nest box correctly placed within the house, the opportunity to forage in a good-sized run or on free range, plus a life free of bullying and stress, will provide optimum conditions for your chickens. But there are some problems that you can encounter. If a bird is straining and not passing an egg, then this is an immediate sign that something is wrong. Pick her up gently and examine her vent area, ideally with an assistant.

Prolapse

This is more likely (although by no means common) in high egg-laying hybrids and older hens. It could be a one-off, but there are reasons for a flock having prolapses, such as excessive artificial daylight hours to encourage early laying, or incorrect nutrition. It is very unpleasant to look at, as part of the chicken's internal reproductive tract is now on the outside and is purple and often swollen.

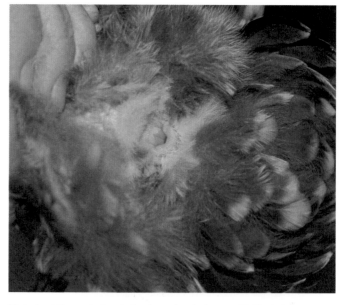

First, check the vent when trying to work out why your bird is straining.

Immediately isolate the hen, as the others will peck the organs, leading to haemorrhage and death. Seek veterinary advice, which will normally involve using a lubricant to gently push back the organs towards the vent. The chicken will also need antibiotics, as her inside organs have been exposed to the outside. There is also a strong possibility that this will not be possible and she will have to be culled.

Egg-Bound

When a hen is egg-bound, she is unable to lay her egg. She will therefore be straining as well as listless and hunched up. It is not as common as is thought, but the cause can be an egg coming the wrong way round due to a calcium imbalance (which has brought about too large an egg). Stress is also thought to be a cause and it is more likely in hens coming into lay. To make things worse, in a high laying hen, more eggs form behind the egg that is jammed.

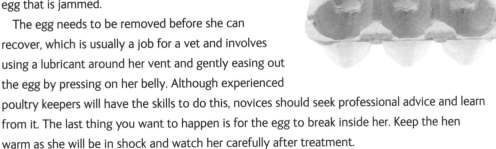

The egg needs to be removed before she can recover, which is usually a job for a vet and involves using a lubricant around her vent and gently easing out the egg by pressing on her belly. Although experienced poultry keepers will have the skills to do this, novices should seek professional advice and learn from it. The last thing you want to happen is for the egg to break inside her. Keep the hen warm as she will be in shock and watch her carefully after treatment.

Egg Peritonitis

This can be mistaken for being egg-bound, though the symptoms will usually be more severe and the abdomen will become swollen. Internally, the egg yolk has fallen into the abdominal cavity (where it shouldn't be), which immediately causes an infection that spreads through the bird's body. It must be very painful. The only solution is to cull humanely as quickly as possible. Egg peritonitis does occur on its own, but can be brought about by rough handling.

Crop and Digestive Problems

The chicken is designed to peck up small morsels of food and grind them with the small, sharp stones lodged in their gizzard. Things can easily go wrong if they peck up the wrong 'feed' or cannot grind.

Prevent Dangerous Substances

Chickens peck at anything, so discarded string is a particular danger, as it can be swallowed and then wraps round inside the crop. Long palatable grass can do the same and grass clippings may also prove dangerous.

Garden chemicals and slug pellets (not that the green poultry keeper will be using these, of course) will also be pecked up. Polystyrene packing balls and similar should be avoided around poultry. Poultry that is closely confined in a run are more likely to peck up dangerous substances and plants through boredom and because they are not able to choose their foraging over a wide area.

Poisonous Plants

Chickens are good at recognizing poisonous plants, but do not test their skills by having any around. Common poisonous plants include:

 The potato and tomato family, which includes Deadly Nightshade
 Foxglove, which produces the heart-slowing drug, Digoxin
 Yew
 Privet and the berries

Laburnum and its related family, such as sweet peas and some vetches

 Some evergreen berries, such as laurel

Crop Impaction (Crop Binding)

The causes, as already mentioned, are usually long grass, string, or other foreign bodies. Occasionally, it can be due to an underlying infection. It shows as a hard lump in the crop and the bird is obviously distressed and unable to eat and may give up trying. It is curable by massaging out the matter, but it must be done by a skilled professional, so seek veterinary advice at once.

Pendulous Crop

Sometimes confused with crop impaction, this is actually due to the muscles of the crop weakening and not being able to channel food into the stomach. The crop hangs down. There are various causes, ranging from the most serious – Mareks disease – to a situation where a bird misses food for a day or so and gorges itself to make up. Treatment depends on cause, but consult a vet.

Diarrhoea

At its simplest, diarrhoea can be caused by a sudden change of diet or even extreme stress. Otherwise, it is likely to be an infection or symptom of a more serious underlying disease. If the bird is otherwise active but loose in its droppings, suspect diet; if the bird is still and hunched up, then isolate, keep warm, give water and feed and consult a vet. It should be noted that bad diarrhoea is a symptom of salmonella infection, which is serious and also can be contagious to humans (pregnant women and the elderly are particularly at risk). Therefore, you must seek veterinary advice.

Lameness in Hens

Hens bustle around being busy and normally keep their leg muscles in good order by foraging and scratching. But they can damage their legs, so watch out for any hopping hens and catch them for a gentle examination.

Cuts, Bruises and Breaks

Broken legs in chickens usually mean humane culling, as they cannot rest them and it is not usually possible to splint. Do not delay, get the chicken humanely despatched. Broken legs are not common, but they do happen. Cut legs could be due to an accident, such as scraping against a wire fence, but do not confuse scaly leg in its bleeding state with cuts. We will deal with scaly legs in our problems section (*see* page 208). Cut legs need cleaning and applying antiseptic or aloe vera gel.

Bumblefoot

This is a more serious problem than it sounds, caused initially by a cut on the underside of the foot. Bacteria gets into the wound, which produces pus and can be painful. It is harder to treat than you might think and you will need veterinary help, as only antibiotics will fully treat the condition. Gently soaking the foot in warm (not hot) water with antiseptic might help to drain the pus, but not always. Make sure that the surfaces of your runs are not sharp, such as pointed gravel, and always remove any sharp objects in the ranging area to avoid this problem.

Notifiable Diseases

Depending on which country you live in, there are diseases that are 'notifiable' to your animal health department. The UK and most countries will include Avian Influenza (bird flu) and Newcastle Disease (once known as fowl pest). Both these are notable in that they show themselves as sudden death and spread among chickens rapidly. So if you have a number of chickens die quickly and at the same time, suspect one of these and consult your veterinary surgeon. Respiratory problems, lack of appetite, and a drop in egg laying are also symptoms.

checklist

Get to know your chickens: Just watching your chickens will help you learn to recognize signs of health and to notice any possible problems early on.

Leave time: Chickens need at least twice-daily attention, so allow yourself time for them.

Seasonal issues: Watch out for extreme heat or freezing cold and take appropriate measures.

Dust in the right place: Don't forget to include a dust bath.

Handling: Learn how to handle your hens for their comfort and safety. Never pick them up or carry them by their legs.

Wing clipping: Decide whether or not your chickens need their flight feathers trimmed.

Don't let them eat everything! Remove or fence off any poisonous plants and don't give chickens long grass or drop string or foreign bodies where they can be pecked up.

Ventilation is key: A well-ventilated but non-draughty poultry house will help to prevent respiratory problems.

Trust in vets: Don't hesitate to call in skilled veterinary advice, especially when you are starting to keep poultry. Even organic poultry flocks sometimes have to use antibiotics.

Reaping Your Rewards

Eggs, Eggs, Eggs

You've done all the hard work and now is the time to enjoy the results – first and foremost, a supply of fresh, tasty eggs. Let's first explain how you get to that bountiful situation.

How Does a Hen Lay an Egg?

The hen is hatched with all the eggs she will ever lay already inside her body, albeit in minute cell form. Correct management and feeding will be needed for them to come to fruition.

The Journey of an Egg

The cell enlarges and becomes an ovarian follicle, which is released into the oviduct, starting at the infundibulum. If it were to be fertilized, this is where this would take place. It then passes into the magnum, where the albumen (the white of the egg) is added. In the isthmus, the shell membrane is added, before passing to the uterus or 'shell gland', where it remains for 20 hours. This is where the shell is formed, plus any colour in the form of pigment. Interestingly, only the blue/green eggs of the Araucana are coloured right through the shell, rather than just on the surface. Once it reaches the vagina, it is laid through the vent, but the vagina extends beyond the vent area to do this, so it is not contaminated with faeces.

How Long Does it Take to Lay an Egg?

From the vagina to expulsion, it is a matter of minutes but, from start to finish, it will take around 25 hours, or slightly less in a hybrid hen. The chicken will then ovulate 30 minutes after laying and the whole cycle begins again. She will not ovulate in the dark and so will miss a day laying from time to time.

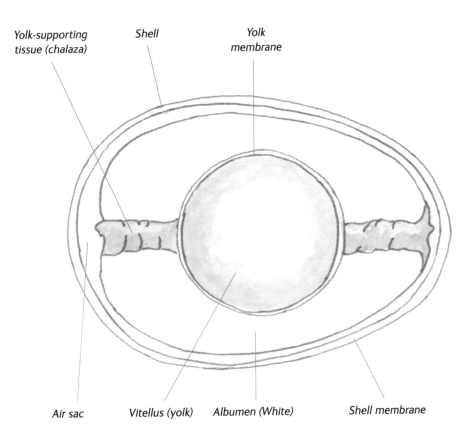

Yolk-supporting tissue (chalaza) Shell Yolk membrane

Air sac Vitellus (yolk) Albumen (White) Shell membrane

What Makes up an Egg?

An egg consists of an inner yolk that is surrounded by albumen, which in turn is bordered by a colourless membrane. The chalazae are two string-like (but much less visible) threads that support the yolk in the albumen. The shell membrane is inside the shell, which is porous, allowing the shell to 'breathe'.

Freshly laid eggs completely fill the shell. The air cell that would be necessary to a developing chick is formed by a loss of moisture through the cell – this is covered in more depth in the incubation section.

Did You Know?

An egg contains roughly 75 calories, depending on its size, so is a great contribution to a healthy low-fat diet.

Egg Facts

- An average egg weighs about 57 g (2 oz)
- The shell forms about 11 per cent of the egg
- The white forms 58 per cent
- The yolk forms 31 per cent
- An egg is about three-quarters water – hence the importance of water in the chickens' management
- Eggs contain vitamins A, B and D, plus minerals such as iron and phosphorus – all essential for human health

A Japanese cockerel stands protectively near his hen, who is incubating a clutch of eggs in a perfectly sized plastic bucket.

Egg Queries

For me, collecting a warm, freshly laid egg never ever loses its charm. But there are a few questions that new poultry keepers often ask about eggs from their chickens.

Does it Matter if They Are Fertile?

If you run a cockerel with your chickens, then the likelihood is that the eggs will be fertilized. Is it all right to eat them? Yes, providing you collect them at least every day and keep them in a cool place – then they will not develop any further and are no different to eat and cook with than unfertilized eggs. What you must not do is to allow a broody hen to sit on them for a few days and then collect them, as they will have started to develop. If you find an abandoned nest of eggs, it is best to discard them.

How Do You Know When Your Hens Are Laying?

You will know because she will tell you, her friends and the whole world. She will cluck loudly and persistently for several minutes. Do try to let her get off the egg before you collect it.

How Big Will My Eggs Be?

Generally, the smaller the bird, the smaller the egg, though there are some exceptions where quite large chickens lay disappointingly small eggs. As the hen first comes into lay, the eggs will be small and sometimes a bit round. As she moves further into her first season's lay, the eggs will increase in size and develop shape. At the start of her lay, she might produce a very tiny egg that is yolkless, or double yolkers (which, if she were breeding, would not be fertile). Then she will settle down to lay a regular supply of eggs of a similar size. In her second season, the eggs may be a bit larger, but she will lay a lower number of them.

Do Misshapen Eggs Signify a Problem?

These do occur, but there is no need to worry unless they happen frequently, in which case it is likely to be a nutritional and, in particular, a calcium problem. Another type of unusual egg is an egg without a shell. Although this may be due to a nutritional defect, again, if it is infrequent, then it's just part of the laying cycle and does happen from time to time.

How to Encourage Good Egg Laying

- Provide well-balanced nutrition – buy a balanced ration
- Make plenty of clean water available during all daylight hours
- Enable hens to forage and/or provide a good-sized run
- Provide an encouraging nest box situated in the darker part of the house, with clean bedding material
- Place a china 'pot egg' in the nest box (or golf balls)
- Provide a stress-free environment, where the hens can lay in safety without fear of aggression or bullying from other hens
- Provide a regular, stress-free routine

Collecting Eggs

This has to be one of the most pleasant things that you can do. It is to be savoured and sometimes shared with good friends or with gentle children. But it is also to be taken seriously and must be done at least daily.

When to Collect

The most convenient time to collect eggs is when you let the chickens out in the morning and again in the evening and that's what most people do. However, if you are at home during the day, collecting at midday when they have had time to lay would be better and, as long as you don't disturb hens laying, the more often you collect the eggs, the less likely the chickens will become broody.

What to Use

Collect eggs in a bowl or clean bucket. Lining with hay or a clean cloth will prevent them getting small cracks that allow in contamination from outside. You shouldn't have dirty eggs but, if you do, put them separately to the clean ones and the same goes for any broken eggs. If an egg is broken in a nest box, clean it out and put in fresh bedding.

Is the Egg Fresh?

The best way to be sure of this is to collect regularly. If you are in doubt, then put the egg in water and, if it sinks, it is fresh; if it floats, it is stale. This is because the air sac in the egg gets larger as the egg gets older, through moisture loss through the shell.

Did You Know?

Can't remember if an egg is fresh or hard boiled? Spin the egg. If it wobbles, it's raw.

Strictly in Order

When you have, say, six hens in full lay, you can easily be getting five or six eggs a day. These quickly mount up. Make sure you store them in order of collecting – that is, newest on the bottom or at the back, so you use the older ones first. Writing the collection date on egg boxes is a useful way to remember.

Keep a Record

Even if it is as simple as writing a number on the calendar every day, do keep a note of how many eggs are being laid throughout the year. It makes for fascinating reading and can also tell you if things are not quite right, as it will give you a pattern. You can also use a special notebook for the purpose.

Storing and Preserving All Those Eggs

There may be times during the year when you have far too many eggs. What can you do to keep them until such time that you have fewer, such as the dark days of winter?

Keep Them Clean

If you keep your housing and runs clean (not muddy and soiled) and pay attention to keeping nest boxes free of droppings or any other soiling, plus you collect once or twice daily, then your eggs should be naturally clean and require no further effort. If they are dirty and you wash them, there is a danger that the dirt on the outside will be taken into the egg in microscopic quantities, as the shell is porous. Use blood-heat water if you have to wash, to close the pores, and invest in a suitable egg wash.

Short-term Storage

Store all eggs point down in a cool but not chilled place out of direct sunlight. A storage bucket does not work because the eggs at the bottom get older and older and sometimes one breaks and makes the others go bad. Use racks or egg boxes and, as we have said, instigate a system where you use the older eggs before you use the freshly laid ones (unless as a very special treat). I know this sounds obvious, but it is less easy than you think when you are bringing in eggs every day and have nowhere to put them.

Long-term Storage

There does not seem to be a way to keep eggs long-term and to still have a product that can be fried or boiled and taste delicious. Stored eggs usually have to be used within a recipe – and well cooked again.

Waterglass

From before the First World War and up until the time when eggs could be bought fresh all year round, waterglass (sodium silicate in solution) was used for storage. The theory is that waterglass will harden the shells and close up the pores, so air cannot get into the eggs. Some people also used to use wax or even a fat such as lard. It is not that easy to get hold of waterglass, as we no longer have high-street agricultural chemists, but some smallholding suppliers have it or you can search the web.

I have to warn that some people who have preserved eggs in this way have not been thrilled with the results and found the eggs to have an unpleasant taste. Remember, when this method was in use, eggs became quite scarce in the winter and they were much more expensive than they are now. Have a go by all means, but do it safely and do not eat the results if you have any doubts about them at all.

Freezing Eggs

These days, we have freezers and this is a good way to store surplus eggs. They should not be frozen hard boiled or in their shells. It is best to break around six eggs at a time into a freezable container, with a pinch of salt or sugar (do label which is which!), then stir very lightly and freeze. Or freeze yolks separately – again, with a pinch of salt or sugar – or whites on their own. The whites can be made into meringues. You can also freeze in ice cube trays – make sure you know how many cubes make one egg, one yolk or one white, so you can calculate for recipes.

Pickled Eggs

A traditional and scrummy way of preserving eggs is to pickle them. It is the vinegar that does the preserving. There are many recipes, but this is a basic method for 'Back-yard Pickled Eggs'.

6 hard-boiled eggs
approx. 600 ml (1 pint) malt vinegar, or sufficient to cover
   the eggs
6 peppercorns
1 whole clove (optional – not everyone likes the taste of these)
2 tsp allspice
fresh dill (again, optional – don't worry if you don't have it)

1: Allow the hard-boiled eggs to cool, then take the shell off the eggs (it's harder with really fresh eggs), discarding any that break or expose the yolk, then place in a sterilized jar (dishwasher will do this) that has an airtight lid.

2: Boil the vinegar and spices together and then simmer for 10–15 minutes.

3: Remove from the heat and allow to cool to about room temperature.

4: You can strain the liquid to remove the whole spices, or leave as is – personal preference.

5: Pour the vinegar over the eggs in the jar, add the dill, if using, close the lid tightly and leave in a cool dark place for a few weeks, after which the eggs will be sufficiently pickled to eat. They will keep for weeks after that but, once opened, do not last long, so eat within a few days.

Storing Cooked Eggs

The final way to store eggs is as part of a final recipe such as in a pudding or pancakes – either savoury or sweet – which you then freeze. You'll be glad you made them when it comes to winter, the snow is on the ground and the days are short. All you will need to do for a taste of the spring and summer will be to defrost your baking and pop it in the oven.

Share Them or Sell Them

The ultimate method of dealing with surplus eggs is to sell them and use the money for chicken feed, so that your household eggs are truly free. If you sell to friends and family, or without grading at your gate, then there are no official regulations, though all sellers have a responsibility to ensure their product is fit for purpose. Of course, you must observe hygiene in collection and it is a good idea to write on the box the day or week they were laid, but not essential. Ask for your boxes back so you can reuse, to avoid another cost. It is quite interesting to keep a note of your sales and balance it up with the cost of chicken feed.

Keep Eggs Safe

Some people have concerns about eating eggs, as they are worried about bacteria within the egg that could have entered via the porous shell or be in the hen's body. This is very unlikely, but for certain groups of people, it is important to cook eggs thoroughly.

At-Risk Groups

Eggs should be cooked *until the whites and yolk are solid* for the following groups, and if you have any doubt at all about the cleanliness of the egg, you might want to wash them and put any washed eggs to one side to also be cooked in this way.

- **The very young – babies to toddlers**
- **The elderly**
- **Pregnant women**
- **Anyone already unwell**

Safe Cooking Practices

Eggs are an ideal medium, when cooked or raw and out of the shell, to harbour bacteria – for example, if someone does not wash their hands and then handles eggs. It is very important to be clean when cooking with eggs. Bacteria can spread easily to other foods, work surfaces or bowls. Here are some top tips for safely cooking with eggs:

- **Keep eggs away from other foods when they are still in the shell and after you have cracked them.**
- **Be careful not to splash egg on to other foods, worktops or dishes.**
- **Always wash and dry your hands thoroughly before and after touching eggs or working with them.**

- Clean surfaces, dishes and utensils thoroughly, using warm soapy water, after working with eggs.
- Don't leave such foods as egg mayonnaise uncovered where it might attract insects or where someone may sneeze on it.

Boiling Eggs

At the end of this section are some recipes to provide you with inspiration in the many ways to cook eggs. But here are the basic times for the classic boiled egg:

Soft-boiled

- **Large eggs**: Cook the eggs for 4 minutes for a runny yolk and 6 minutes for a medium-boiled egg, where the yolk is only slightly runny.
- **Medium-sized eggs**: Reduce the above cooking times by one minute.
- **Bantam eggs**: These cook quite quickly, so reduce the time even further.
- **Extra-large eggs**: Add on one minute to the large-egg cooking time.

Hard-boiled

For large eggs, you need to cook them for 8 minutes or longer and, for all other sizes, reduce or increase the amount to the same ratios as given above for soft-boiled eggs.

Chickens for the Table

For some back-yard flocks, where the keeper is breeding birds, one of the benefits is a table bird from time to time, when there are surplus cockerels. For others, some cockerels might be deliberately reared for meat.

Why Rear Table Birds?

Arguably, the overall reason is that you can be assured of their welfare all the way through their lives to their deaths and take control of their food. This is a big responsibility and not to be taken lightly. If you choose a commercial table broiler hybrid to rear, their life is short – around eight to ten weeks – and you need to ensure that it has a well-managed and comfortable existence during that time. That means the

opportunity to be able to move freely (a large run or safe access to free range) and to be able to forage at least part of the time. Of course, access to suitable feed, water and shelter and protection from predators should be seen as a standard necessity with any bird.

Humane Slaughter

In addition to this, you must be able to bring about a quick and humane death. It is not easy to kill any bird and not a skill to be learnt from a book, but from someone who is an expert – and even then, it should be practised on dead chickens first. The Humane Slaughter Association, who offer excellent online advice and have a series of technical leaflets, prefer pre-stunning but, for back-yard slaughter, they advise neck dislocation by a skilled person, without using any gadgets, such as pliers or similar. This is considered the best method of quick and humane slaughter.

'Humane' Includes the Handling Prior to the Kill

Catching is all part of the slaughter process, so never chase chickens round prior to slaughter, but carry it out at night by gently removing the bird from its roost and, with the watch-word being quickly, despatch it immediately. It might be stressful for you, but your job is to make it as stress-free as humanely possible for the bird, so plan ahead. Check out the Humane Slaughter Association website for more details: www.hsa.org.uk.

Plucking and Drawing

Plucking is done either wet or dry. Wet plucking is as it sounds – dunk the chicken in hot water first – some people believe it makes the bird much easier to pluck. Dry plucking is also self-explanatory – just hang the dead bird up and start pulling out feathers until they are all gone.

Practise, Practise, Practise

It is possible to get small-scale plucking machines, but you really need quite a few birds to justify one. If you have never plucked before, then please do so before you undertake any number of birds. It takes ages when you have never done it before, while an expert can do it in around 15 minutes. You will take rather longer to begin with, but practice makes perfect! That is why it really is best to start small and not undertake to kill or pluck more than half a dozen to start with.

Once Plucked

You will also need to gut (the 'drawing' part) and to be able to 'present' the carcass (cutting off the feet and neck, trussing) – check out cookbooks for ways in which to do this. Keep the giblets and neck and put in a separate bag to accompany each bird, so that gravy can be made using them. It is really best to plan to keep your first attempt at table chickens for your own freezer until you have perfected plucking and gutting and carcass presentation.

Timeline for the Table

If you want to rear birds for the table, this should be the order of events (see pages 246–50 for further information on rearing chicks):

- **Phone supplier and order chicks, bearing in mind the intended slaughter date.**
- **Order chick crumbs, grower ration and finisher ration.**
- **Get rearing equipment prepared and in place, including safe water containers. An hour or so prior to delivery or collection, switch on heat lamp.**
- **Rear chicks under heat until they have feathered, gradually reducing heat.**
- **Change to a growers ration and encourage exercise.**
- **Put on to finishing pellets for final two weeks.**
- **Plan humane slaughter and plucking and carry out quickly and calmly.**
- **Make sure you have enough freezer space available.**

As regards reducing heat, do this according to the weather outside – if rearing for Christmas, the temperatures outside could be very low. Reduce heat by gradually raising the heat lamp and watching the chicks to see if they are huddling (too cold) or spread out (too hot). Perhaps move outside on warm sunny days (with shelter) and put back with heat at night. Finally, remove heat altogether but keep watching for any signs of distress. Note – make sure that the rearing pen is draught free.

Cooking Cockerels, Bantams and Older Birds

For many chicken keepers, the only table birds will be the odd surplus cockerel, and I have found that it is much better to cook these (young or old, any breed that isn't a dual-purpose type) as if they were game birds rather than as for chickens. Some bantams also taste a bit 'gamey', which is very flavoursome, so cook them as for pheasants. Older birds require longer cooking – casserole, don't roast.

Eating Laying Hens

Very few chicken keepers eat their old laying hens, as most tend to keep them until they are past their best. However, if you do decide to eat a laying hen, remember that she will have a very light frame indeed, so perhaps remove the breasts and discard the rest of the body (this avoids plucking). If you want to use the whole bird, she will need to be casseroled or curried. Alternatively, you could use everything but the breasts for a stock.

Recipes: Eggciting Ways With Eggs

When you have a bowl of lovely fresh eggs and plenty of them, then it is time to think of some new ways with eggs. An egg is such a versatile food and can be eaten savoury or sweet and in so many ways that it is worth researching some recipes. In this section, we have found some exciting recipes for you to try – for eggs mostly, but a couple for meat too – but don't be frightened to try your own or simply use them as inspiration.

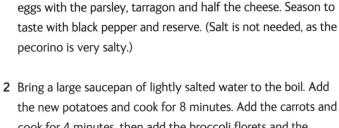

Vegetable Frittata

Serves 2

6 eggs
2 tbsp freshly chopped parsley
1 tbsp freshly chopped tarragon
25 g/1 oz/¼ cup finely grated
  pecorino or Parmesan cheese
freshly ground black pepper
175 g/6 oz/1¼ cups tiny
  new potatoes
2 small carrots, peeled and sliced
125 g/4 oz/1¼ cups broccoli,
  cut into small florets
1 courgette/zucchini, about
  125 g/4 oz/¾ cup, sliced
2 tbsp olive oil
4 spring onions/scallions,
  trimmed and thinly sliced

To serve:
mixed green salad
crusty Italian bread

1 Preheat the grill/broiler just before cooking. Lightly beat the eggs with the parsley, tarragon and half the cheese. Season to taste with black pepper and reserve. (Salt is not needed, as the pecorino is very salty.)

2 Bring a large saucepan of lightly salted water to the boil. Add the new potatoes and cook for 8 minutes. Add the carrots and cook for 4 minutes, then add the broccoli florets and the courgette/zucchini and cook for a further 3–4 minutes, or until all the vegetables are barely tender. Drain well.

3 Heat the oil in a 20.5 cm/8 inch heavy-based frying pan. Add the spring onions/scallions and cook for 3–4 minutes until softened. Add all the vegetables and cook for a few seconds, then pour in the beaten egg mixture. Stir gently for about a minute. Cook for a further 1–2 minutes until the bottom of the frittata is set and golden brown.

4 Place the pan under a hot grill for 1 minute, or until almost set and just beginning to brown. Sprinkle with the remaining cheese and grill for a further 1 minute, or until it is lightly browned. Loosen the edges and slide out of the pan. Cut into wedges and serve hot or warm with a mixed green salad and crusty Italian bread.

Stilton, Tomato & Courgette Quiche

Serves 4

For the shortcrust pastry/piecrust:

225 g/8 oz/2 cups plain/
  all-purpose white flour
pinch salt
50 g/2 oz/4 tbsp white vegetable
  fat/shortening or lard
50 g/2 oz/½ stick butter
  or block margarine

For the filling:

25 g/1 oz/2 tbsp butter
1 onion, peeled and
  finely chopped
1 courgette/zucchini, trimmed
  and sliced
125 g/4 oz/1 cup Stilton/blue
  cheese, crumbled
6 cherry tomatoes, halved
2 large/extra-large eggs, beaten
200 ml/7 fl oz/¾ cups crème
  fraîche/sour cream
salt and freshly ground
  black pepper

1 Sift the flour and salt into a mixing bowl. Cut the fats into small pieces and add to the bowl. Rub the fats into the flour using your fingertips until the mixture resembles fine breadcrumbs. Add 1–2 tablespoons cold water and mix to form a soft, pliable dough. Knead gently on a lightly floured surface until smooth and free from cracks, then wrap and chill for 30 minutes.

2 Preheat the oven to 190°C/375°F/Gas Mark 5. On a lightly floured surface, roll out the dough and use to line an 18 cm/ 7 in lightly oiled flan tin/tart pan, trimming any excess dough with a knife.

3 Prick the base all over with a fork and bake blind in the preheated oven for 15 minutes. Remove the pastry from the oven and brush with a little of the beaten egg. Return to the oven for a further 5 minutes.

4 Heat the butter in a frying pan and fry the onion and courgette/zucchini for about 4 minutes until soft and starting to brown. Transfer into the pastry case/tart shell. Sprinkle the Stilton/blue cheese over evenly and top with the halved cherry tomatoes.

5 Beat together the eggs and crème fraîche/sour cream and season to taste with salt and pepper. Pour into the pastry case and bake in the oven for 35–40 minutes until the filling is golden brown and set in the centre. Serve the quiche hot or cold.

Aromatic Eggs

Serves 6 as a stsrter

2 tbsp jasmine tea leaves
9 eggs
2 tsp salt
4 tbsp dark soy sauce
1 tbsp soft dark brown sugar
2 whole star anise
1 cinnamon stick
2 tbsp sherry vinegar
2 tbsp Chinese rice wine or
   dry sherry
2 tbsp caster/superfine sugar
$\frac{1}{4}$ tsp Chinese five-spice powder
$\frac{1}{4}$ tsp cornflour/cornstarch

Top Tip

This recipe is ideal for bantam eggs. Allowing 12–18 eggs to serve 6 people, depending on size, simmer for 2 minutes in step 2, and for a further 2 minutes in step 3. Leave to soak and peel as before. Leave whole, if small enough.

1 Place the tea leaves in a jug and pour over 150 ml/$\frac{1}{4}$ pint/$\frac{2}{3}$ cup boiling water. Leave to stand for 5 minutes, then strain, reserving the tea and discarding the leaves.

2 Meanwhile, place the eggs in a saucepan with just enough cold water to cover them. Bring to the boil and simmer for 4 minutes. Using a slotted spoon, remove the eggs and roll them gently to just crack the shells all over.

3 Add the salt, 2 tablespoons of the soy sauce, the dark brown sugar, star anise and cinnamon stick to the egg cooking water and pour in the tea. Bring to the boil, return the eggs to the saucepan and simmer for 4 minutes. Remove from the heat and leave the eggs for 2 minutes, then remove the eggs and plunge them into cold water. Leave the tea mixture to cool.

4 Return the eggs to the cooled tea mixture, leave for 30 minutes, then drain and remove the shells to reveal the marbling.

5 Pour the remaining soy sauce, the vinegar and Chinese rice wine or sherry into a small saucepan and add the caster/superfine sugar and Chinese five-spice powder. Blend the cornflour/cornstarch with 1 tablespoon cold water and stir into the soy sauce mixture. Heat until boiling and slightly thickened, stirring continuously. Leave to cool.

6 Pour the sauce into a small serving dish. Cut the eggs widthways into quarters, place in a serving bowl or divide between individual plates and serve with the dipping sauce.

Chinese Egg Fried Rice

Serves 4

250 g/9 oz/1²/₃ cups
  long-grain rice
1 tbsp dark sesame oil
2 large/extra-large eggs
1 tbsp sunflower oil
2 garlic cloves, peeled
  and crushed
2.5 cm/1 in piece fresh root
  ginger, peeled and grated
1 carrot, peeled and cut
  into matchsticks
125 g/4 oz/2 cups
  mangetout/snow peas, halved
220 g/8 oz can water chestnuts,
  drained and halved
1 yellow pepper, deseeded
  and diced
4 spring onions/scallions,
  trimmed and finely shredded
2 tbsp light soy sauce
¹/₂ tsp paprika
salt and freshly ground
  black pepper

1 Bring a saucepan of lightly salted water to the boil, add the rice and cook for 15 minutes or according to the packet instructions. Drain and leave to cool.

2 Heat a wok or large frying pan and add the sesame oil. Beat the eggs in a small bowl and pour into the hot wok. Using a fork, draw the egg in from the sides of the pan to the centre until it sets, then turn over and cook the other side. When set and golden, turn out on to a board. Leave to cool, then cut into very thin strips.

3 Wipe the wok clean with absorbent kitchen paper, return to the heat and add the sunflower oil. When hot, add the garlic and ginger and stir-fry for 30 seconds. Add the remaining vegetables and continue to stir-fry for 3–4 minutes until tender but still crisp.

4 Stir the reserved cooked rice into the wok with the soy sauce and paprika and season to taste with salt and pepper. Fold in the cooked egg strips and heat through. Tip into a warmed serving dish and serve immediately.

Egg Custard Tart

Serves 6

For the sweet pastry:
50 g/2 oz/½ stick butter
50 g/2 oz/¼ cup white
  vegetable fat/shortening
175 g/6 oz/1⅓ cups plain/
  all-purpose flour
1 egg yolk, beaten
2 tsp caster/superfine sugar

For the filling:
300 ml/½ pint/1¼ cups milk
2 eggs, plus 1 egg yolk
2 tbsp caster/superfine sugar
½ tsp freshly grated nutmeg

1 Preheat the oven to 200°C/400°F/Gas Mark 6, 15 minutes before baking. Oil a 20.5 cm/8 in flan tin/tart pan. Make the pastry by cutting the butter and vegetable fat/shortening into small cubes. Add to the flour in a large bowl and rub in until the mixture resembles fine breadcrumbs. Add the egg yolk, sugar and enough water to form a soft and pliable dough. Turn on to a lightly floured surface and knead. Wrap and chill in the refrigerator for 30 minutes.

2 Roll the dough out on to a lightly floured surface and use to line the oiled flan tin. Place in the refrigerator to chill.

3 Warm the milk in a small saucepan. Briskly whisk together the eggs, egg yolk and sugar. Pour the milk into the egg mixture and whisk until blended. Strain through a sieve into the pastry case/pie crust. Place the flan tin on a baking sheet.

4 Sprinkle the top of the tart with nutmeg and bake in the preheated oven for about 15 minutes. Turn the oven down to 170°C/325°F/Gas Mark 3 and bake for a further 30 minutes, or until the custard has set. Serve hot or cold.

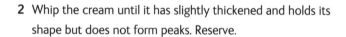

Spicy White Chocolate Mousse

Serves 4-6

6 cardamom pods
125 ml/4 fl oz/¹/₂ cup milk
3 bay leaves
200 g/7 oz white chocolate
300 ml/¹/₂ pint/1¹/₄ cups
   double/heavy cream
3 medium/large egg whites
1–2 tsp cocoa powder, sifted,
   for dusting

1 Tap the cardamom pods lightly so they split. Remove the seeds, then crush lightly in a pestle and mortar. Pour the milk into a small saucepan and add the crushed seeds and the bay leaves. Bring to the boil gently over a medium heat. Remove from the heat, cover and leave in a warm place for at least 30 minutes to infuse. Break the chocolate into small pieces and place in a heatproof bowl set over a saucepan of gently simmering water. Ensure the water is not touching the base of the bowl. When the chocolate has melted, remove the bowl from the heat and stir until smooth.

2 Whip the cream until it has slightly thickened and holds its shape but does not form peaks. Reserve.

3 Whisk the egg whites in a clean, grease-free bowl until stiff and standing in soft peaks.

4 Strain the milk through a sieve into the cooled melted chocolate and beat until smooth. Spoon the chocolate mixture into the egg whites, then, using a large metal spoon, fold gently. Add the whipped cream and fold in gently.

5 Spoon into a large serving dish or small individual cups. Chill in the refrigerator for 3–4 hours. Just before serving, dust with a little sifted cocoa powder.

Chicken Marengo Casserole

Serves 4

4 chicken portions, skinned
1 tbsp olive oil
15 g/½ oz/1 tbsp
   unsalted butter
1 onion, peeled and cut
   into wedges
2–3 garlic cloves, peeled
   and sliced
2 tbsp plain/all-purpose flour
900 ml/1½ pints/1 scant quart
   chicken stock
3–4 small tomatoes, peeled
salt and freshly ground
   black pepper
1 fresh bay leaf
350 g/12 oz/3½ cups new
   potatoes, scrubbed and
   cut in half
75 g/3 oz/½ cup
   sweetcorn kernels
350 g/12 oz/11½ cups
   fresh spinach

1 Preheat the oven to 180°C/350°F/Gas Mark 4. Lightly rinse the chicken and pat dry on absorbent paper towels.

2 Heat the oil and butter in an ovenproof casserole (or frying pan, if preferred), add the chicken portions and cook until browned all over. Remove with a slotted spoon and reserve.

3 Add the onion and garlic and cook gently for 5 minutes, stirring occasionally. Sprinkle in the flour and cook for 2 minutes before stirring in the stock and bringing to the boil.

4 If a frying pan has been used, transfer everything to a casserole, and return the chicken to the casserole dish with the peeled tomatoes. Season to taste with salt and pepper and add the bay leaf. Cover with a lid and cook in the oven for 30 minutes. Remove the casserole from the oven and add the potatoes and sweetcorn. Return to the oven and cook for 30 minutes. Add the spinach and stir gently through the casserole. Return to the oven and cook for a further 10 minutes, or until the spinach has wilted. Serve.

Pests,
Predators
& Problems

Parasites

Prevention is better than cure. Good management, observation and time will greatly help to keep your chickens healthy, but there are also some special preventative measures for the challenges that parasites present.

External Parasites

These creepy crawlies come in a range of types and are much more dangerous to the chickens' health than merely making them very, very uncomfortable. The commercial poultry farmer spends a lot of time and money ensuring that his flock does not suffer from them, and the green back-yard poultry keeper must also have a preventative plan and not wait to find them on the chickens.

External Parasite: Beware Red Mite

Red mite is the best known of the chicken mites. It is a bloodsucker that lives in cracks and crevices of the poultry house during the day and seems to have an affinity with wood. It can also live in trees whose branches grow over the run. During the night, it comes out and feeds voraciously on the roosting chickens.

It is essential to check the ends of perches for red mite infestation.

Why Red Mite?

Red mite has been a problem with poultry keepers, large and small, for years, but we have changed our method of poultry keeping since the war. Before, and still after, the war, it was an 'all in, all out' system, with the house creosoted (now restricted in use in many areas) and the ground limed and left for a few weeks. Thus the mite was always controlled. Now we tend to

keep our chickens until they die, or certainly more than one season, so the mite gets the opportunity to really take a hold.

Wild birds can also carry red mite. Unless you shut your chickens in a covered run, it is almost impossible to stop wild birds and your chickens coming into contact, so it's best just to accept there is a risk and take the preventative measures as described above (*see page 207*).

Identifying Red Mite

The most unpleasant way to discover that you have red mite is to find dead chickens in the house. Although tiny, red mites are devastating in numbers – and they breed quickly. Their bloodsucking activities cause anaemia, which can and does result in death, especially in older birds, those who are laying every day and young birds or chicks. Keep an eye out for:

 Itching and lesions: If you are handling and examining your chickens every week, probably the first symptom you will actually notice is that, when you put them down, *you* are very itchy indeed. There may well be lesions on the breast or legs of the birds as well. The chicken will be very itchy and will scratch at herself.

 Blood-filled insects: Inside the house, they look like grey particles of dust and in the morning they will be easy to spot, as they will be full of blood and show as bright red specks, mainly at the ends of or underneath the perches, but also in the crevices of the house. Use a kitchen towel and, as you wipe them, the blood will be released and the towel will have red flecks.

Lethargy: You may also notice that the chickens are lethargic in the mornings (when they are anaemic) and perk up in the afternoon after they have fed, before going back to roost to be sucked again by the mite.

Fewer eggs: Egg production will take a steep drop as a result of mite.

Keeping away from the source: Another symptom is the reluctance of hens to go into the hen house at night, preferring to roost outside.

Getting Rid of Red Mite

This is much easier said than done. The life cycle of the mite can be complete in seven days, so they breed at a remarkable rate. Miss a few and they will quickly restore their numbers. So the most important thing is to undertake treatment extremely thoroughly.

Treat the birds: Remove the birds and treat according to your veterinary advice or with a product purchased from an agricultural merchant. There are an increasing number of products on the market now that are specific to dealing with this challenging mite. If you have a big infestation, do contact your vet, as there are products that work well but are not strictly licensed for poultry and need a veterinary prescription. They are very effective.

Treat the house: Do this at the same time as treating the chickens. If you do not treat the house, then the mite will soon re-establish themselves on the chickens. Normal disinfectants and detergents may help to shift them, but will not kill them all or their eggs, so choose a product that is made for washing out mite-infested poultry houses. You may need to wash out the house a couple of times and really get into the cracks and crevices. It will have to be thoroughly swept out first, so there are no hiding places. Leave it open for the sunlight to reach, as mites do not like sunshine.

The organic option: There are organic mite treatments on the market and, if you choose this option, it is absolutely vital that you wash the house out as already mentioned and use them as indicated on the label. The same goes for products that go on the birds. For the chickens' welfare, you have to get rid of these mites. A blowtorch (use carefully on wood) can also be used in the crevices and is very effective.

Prevention of Mite is Essential

Mites are at their most active in the summer, but mite prevention should be undertaken all year round. Do the following to prevent, not just to treat an infestation:

- **Clean and treat houses**: Keep chicken houses clean and make plans to regularly treat as routine for mite – if you do this all through the summer and early autumn months, you may be able to resume in the spring and leave the winter treatment.

- **Treat and inspect chickens**: Routinely treat chickens for mite all year round and inspect them carefully for signs.

- **Where will you put the chickens?** Have an alternative place you can put the chickens for a few days while you thoroughly treat their house and run.

Did You Know?

Red mite can live for many years without a host (chickens), so if you buy a house that has had chickens in it, treat it thoroughly for red mite before you use it.

Red Mite and Humans (and Other Animals!)

Red mites will happily go on to humans or pets. It feels like a wave of feathers hitting your face and then you begin to itch. It is difficult to see, but you can feel it. At this point, having already started to think about how to get rid of it, make sure you shower and wash your clothes before you sit in a chair. Remember, red mites love to live in cracks and crevices and a sofa in your nice warm house is as good as a hen house to them. They will also get on household pets such as cats and dogs and even ponies, who will also need treating if this is the case.

External Parasite: Scaly Leg Mite

This is a seriously unpleasant microscopic mite that goes by the scientific name of *Knemidocoptes mutans*. It burrows under the scales of the legs, hence the name, and irritates them in doing so, which causes the scales to rise and a crust to form. That unpleasant and lumpy-looking crust is a mixture of the mites' excreta and skin flakes, and has a very irritated and itchy leg underneath. It may also bleed and, in extreme cases, the chicken is lame.

Susceptible Breeds

Although all chickens could get scaly leg, the ones with clean, tight-scaled legs are not so much at risk. It is the feathered-leg breeds that suffer so badly, so you need to be especially vigilant if you keep any of these. Unscrupulous sellers of these breeds have been heard to tell new buyers that 'these type of breeds always have scaly legs'. Well, they should not be allowed to develop it, and there are plenty of treatments for the condition, so there is no excuse for any bird to have scaly leg.

Treating Scaly Leg

Again, prevention is better than cure, especially when you know a breed is particularly prone to attracting these mites, but there are now a range of products on the market that will effectively deal with scaly leg mite. It will take time for the birds' legs to return to normal and, in bad cases, they may remain deformed but at least mite-free.

- **Treat the birds:** Choose one from your local agricultural merchant or consult your vet, who may be able to prescribe a topical treatment such as the single spot treatment (i.e. applied in a single drop) used on household pets.

- **Treat the house:** The chicken house will need treating to stop any further spread of the mite.

- **Older treatments:** These included softening the crusts by applying petroleum jelly, which was also thought to act as a medium to suffocate the mites. The legs were then scrubbed in a medicated but not strong shampoo, using a soft toothbrush. You could use an organic variety. Be gentle – do not draw blood. The legs were also immersed for up to a minute in surgical spirit before being scrubbed but, as this is very abrasive, they should not be bleeding when you do this.

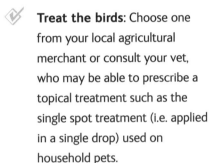

There were some harsher old treatments, but anything that is going to cause the bird discomfort should not be used, which is why modern sprays, which are effective and easy to use, are usually the best way to treat this invasive mite.

More Mites ...

Northern Fowl Mite

This is related to the red mite but it is less common and, unlike that one, it stays on the bird all the time. It can often be seen round the vent and, in light-coloured birds, it looks like some dirt around the bottom. It loves to live in the crests of crested breeds. The other symptoms are the same as red mite: the birds become anaemic and may die if not treated and, in the earlier stages, are itchy, lethargic and have a drop in egg production. The treatment is also the same as for red mite.

Depluming Mite

This is not seen so often. As its name suggests, it damages the feathers, so is particularly unwelcome in birds that are going to be shown. It is dealt with in the same way as the usual mites.

External Parasite: Lice

These are relatively easy to spot under the wings and round the vent, and they move to get out of daylight, so you can see them scuttle. They deposit their eggs around the feather shafts, which are visible. You can try removing these with your fingers and burning them. Lice, unlike mites, feed off skin and feather debris rather than suck the blood, but they still irritate the bird – and you if you handle it. Purchase a product for controlling lice and use it regularly, as directed.

Internal Parasites

Internal parasites are usually worms. Poultry can be infected by a number of worms, including those that affect household pets and wild birds. Again – get there before the worms do! It is best to have a routine worming programme – consult your vet for your local situation.

Roundworms

Also known as 'ascarids', these are the most frequently found worms – they are easily transmitted chicken to chicken or by a third-party host. They live in the intestine. Symptoms include a drop in egg production, weight loss and loose droppings, as the bird's digestive system struggles to cope with the burden of having nutrients drained from it by invasive worms. Treat with a wormer prescribed by your vet. Moving to clean pasture or moving the run must also be part of the preventive process or re-infection will occur.

Tapeworms

These are the segmented, flat, ribbon-type worms – you might see them in droppings. They don't seem to affect poultry as much as roundworms, but they should be eradicated from your premises anyway, and this is done by an all-purpose worming medication prescribed by your vet.

Top Tip – Withdrawal Periods

It is very important that you follow the 'withdrawal periods' given with the worming medicines and discard eggs during this time. Table birds should not be killed for meat until the withdrawal period has passed.

Gapeworm

A particularly unpleasant worm that lives in the throat (trachea) and causes the bird to gape – that is, to stretch the neck upwards and open and close the beak. This is almost like a gasping action, although the breathing is not always affected. The bird is trying to get rid of the worms. Gapeworms are present in earthworms, which in turn are ingested by the bird. Prevent by rotation of your chickens and allowing plenty of space per chicken – do not overcrowd. Use a wormer prescribed by your vet.

Coccidiosis

This is caused not by a worm but by a protozoal parasite that lives and breeds in the gut and is excreted as 'oocysts' in the droppings, where other birds pick it up. It is therefore easily transmitted bird to bird and flourishes in wet, dirty litter, in drinkers and feeders. It is usually seen in young birds or young adults, especially when batches are constantly reared in the same pen. Therefore, coccidiosis should be prevented by cleanliness, rotation of housing and runs and, of course, not overcrowding. But it does occur sometimes even in well-managed flocks, as it multiplies so quickly.

Symptoms include dullness, inability to grow, diarrhoea, appearing cold, huddling, hunching and generally being uncomfortable. There may be some deaths. Consult your vet for medication; he or she may also check droppings to be sure that coccidiosis is the cause. Starter and grower rations can include an anti-coccidiostat, which will help to avoid the condition, but you should pay attention to prevention regardless of whether or not you use these feeds.

Safety and Security

As we have seen, chickens are at risk from many predators who want to do them harm or eat them. It is your responsibility to keep your stock safe and sound from these predators, which come in many forms: winged, four-legged and even sometimes two-legged.

The Fox

Foxes are perhaps top of the poultry keeper's list when it comes to predators. In theory, they are nocturnal hunters, so if you shut in your poultry at dusk, they should be safe. In practice, foxes can and do hunt in the daytime and in towns and cities, they seem to be awake all day long. Also, when the time of the year during which the female has growing cubs coincides with short, dark nights, hunger – the need for her to feed her family as well as herself – means they come out in the day for more chances of food.

Keep Them Out

The best method for most poultry keepers is prevention, which is to keep poultry confined in fox-proof houses and strongly fenced runs except when there is someone outside with them, such as weekends spent in the garden. Always make sure that the poultry are shut up when they go in to roost at dusk, even if that means asking neighbours.

Can You Get Rid of Foxes?

There are many obvious barriers – legal, moral, social and practical – to lethally disposing of foxes. Successful disposal is a job for an expert marksman, and trapping still involves killing the animal, which must be done humanely. In any case, removing a fox from the area in any way is likely to be done in vain – foxes are territorial, so as soon as you get rid of one, another one will move in.

Deterrence

The presence of dogs can help to deter foxes (though *see* pages 39 and 215). Also, it is well worth trying some of the chemical and ultrasonic repellents that are on the market.

The Weasel Family

Weasels and stoats are small, so can squeeze through any tiny gaps into the house, where they will wreak havoc and cause deaths. Be sure that the house is secure, and check regularly. Mink can be a problem in some areas. They usually kill at night, so make sure the poultry are securely shut away. Mink normally only live near water.

Other Four-legged Predators

These include raccoons, polecats (and escaped ferrets), bobcats and badgers who are looking for eggs. As with all the others, protect your poultry with secure housing, meticulous attention to shutting in at night and well-constructed runs.

Winged Predators

Winged predators in the form of certain types of hawk, owls, crows, jays and magpies, are capable of harming birds and young chickens – chicks and small bantams are particularly at risk. Most varieties of raptor are protected and cannot be killed, so protect your poultry. If you have a big problem with this, you may have to net the run.

Domestic Danger

Chickens do suffer from attacks by domestic dogs. As well as protecting your chickens from the attacks, you will also need to seek out the owner and, if necessary, take legal advice. Domestic cats are not usually a threat to fully-grown chickens that should already be well-protected from all predators.

'Guard Pets'

Your own family cat will keep away other cats, as will your dog. Your own dog is good protection not only against other dogs but also against foxes, who do not normally like to come where there is a loose dog – though with the semi-tame foxes, this is not always the case.

Do Not Become Complacent

However, sadly, it is often people's own family dog that can do the most damage. You must train your dog at the very beginning not to touch the birds and never, ever allow them to chase them, even if they do not kill them. If you cannot train your dog, then it must be kept away from the chickens and only ever be close to them if on a lead. Do not let friends' or family members' dogs loose unless you are certain they are well behaved. I have lost a much-loved chicken to a friend's badly behaved dog, having been assured that he wouldn't do any harm. Remember, too, that your poultry will naturally be afraid of dogs, so even with a harmless family pet, be careful that he does not inadvertently frighten them.

Top Tip

If you have a predator attack, then learn from it. Identify which predator it was and adapt your housing and management accordingly. Do not let it happen again.

Rats and Mice

It is always said that poultry attract vermin but, although this tends to be true, you don't have to let them move in. Be on the lookout for the first signs of unwanted visitors and take action.

Discouraging Vermin

Keeping your chickens and their run clean and having an efficient and orderly compost heap will help to discourage rats and mice. They come where the food is, so, although your poultry will need feed available, do not overfeed – you do not want waste food all over the run and house. Use suitable feeders

Hanging your daily food rations helps keep them away from vermin.

and if there is a particular problem in your area, use an auto-feeder that will discourage them. Store all your feed in tight-lidded metal bins, not sacks.

I See a Rat ...

If you see one rat, you can be sure there is another close by. A rat has quite a large territory and you need to control the numbers before they explode. There are several methods: shooting, trapping and poisoning. All must be carried out responsibly, that is to say, with the least amount of discomfort to the rat.

Shooting must be done by someone who can kill the animal outright; if you choose to trap, you must check the traps every day and have a plan as to how to dispose of the live rat; poisons have become quicker and more reliable, but the bait must be safely set away from domestic animals and children and the resulting corpses quickly cleared (there is nothing a chicken likes better than a mouse to eat, so it is very important that they do not eat a poisoned one); and drowning is NOT an option.

Odd Behaviour

From time to time, the behaviour of chickens can cause some concern – common strange behaviours being egg eating and pecking at other hens. Nearly always, it is due to not understanding their natural needs and therefore not meeting them.

Egg Eating

This is a difficult behaviour to stop once it has started, so it's important to gear your management towards preventing this annoying and also devastating behavioural problem. It is devastating because a hen can encourage others to do it as well and the worst case scenario is that you have to cull all the hens and start again or else devise methods so that the eggs cannot be pecked once laid.

What is It?

Egg eating is when one or more hens peck at their own or their flock mates' eggs until they break them and then they eat them. It is horrible to watch and it must be dealt with immediately, before the situation worsens.

Preventing Egg Eating

Usually (but not always, there are exceptions) correct management and understanding how hens behave will discourage any form of egg eating. Overcrowding, boredom and irregular collection of eggs will inevitably lead to this behaviour occurring. It is important to understand that it is natural for hens to peck at things and eggs left uncollected are an obvious target. Nest boxes too should be private, situated in the darkest place in the house or designed so that, once the hen has laid, the eggs are not on show to all the other hens.

How to Stop Egg Eating

Firstly, identify why it has happened and correct the management problem. If hens are bored and overcrowded, enlarge the run, give them something to do – hang up vegetables or seed blocks for them to peck at. Try to let them free range for at least part of the day or even week. Then look at the nest boxes and take whatever action is necessary to make them secure from the casual attentions of the other hens. Other actions you can take include:

- **Separation:** Remove the culprit to another house with perhaps one other hen.

- **Decoys:** Place pot eggs (china eggs) in her nest so that pecking at them does not produce anything.

- **Collection:** Be sure to collect this hen's eggs as soon as you can after laying.

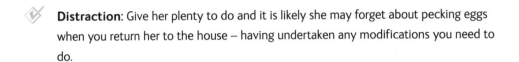

Distraction: Give her plenty to do and it is likely she may forget about pecking eggs when you return her to the house – having undertaken any modifications you need to do.

An old cure: Fill an egg full of mustard so that pecking at it is an unpleasant experience for the chicken. No one has recorded how successful this is!

Confiscation: If you really cannot break the habit, it is possible now to buy internal, freestanding nest boxes with a 'roll away' tray. This means every egg that is laid rolls away from the nest box and is covered with a plastic sheet so the hen cannot get to it.

Pecking Other Hens

Assuming that you have not introduced new hens, especially those of a different age, into the group, this is quite a tricky behavioural problem to solve.

Prevention of Pecking

It all comes back to management again. Bored hens in too small a space with nothing to do will find something, and that is likely to be pecking each other. Also, the ones lower down the pecking order will not be able to get out of the way and will become easy targets for those higher up, irritated by them, as they are in their space.

Take Immediate Action

It is your duty of care as a poultry owner to take care of all your birds and that includes the weakest. What you must never do is stand back and think, oh well, they will work it out eventually. If a bird is pecking another and drawing blood, you must remove either the one being pecked or the pecker. Or both. You need to disrupt their environment temporarily.

Be Honest – Is it the Management?

Is the run bare, are the birds unable to get away from each other because the space is not big enough, is there something for the hens to do on a daily basis? If not, then take steps to rectify this before you return the bird or birds you have removed. Otherwise, the same situation will occur all over again.

Treating the Victim

As well as having painful wounds that must be treated with antiseptic spray, the victim is likely to be losing weight because she dare not eat in the other bird's presence – and to be dehydrated because she dare not drink. Provide a vitamin supplement in the water and plenty of nutritious food to build her up again. She cannot be returned until her wounds have healed, as chickens will always peck at blood and it is likely the rest of the flock will join in. It is possible for a chicken to be pecked to death.

Returning to the Flock

When her wounds have healed, you can return her (and the one pecking her, if you have also removed her – keep them separately but near to each other) to the flock. By the time you have done this, the pecking order will have changed and it's likely that the one doing the pecking will have gone down the line. Watch them very carefully. If you have changed the management, then the problem should end here. If she is still bullied, you will have to keep her, and perhaps a less aggressive hen as a companion, separately.

Let's Hear it for the Boys!

As we have seen, it is not necessary to keep a cockerel to get eggs but, if you do, then there is a possibility that his crowing will draw comments from neighbours. Can you stop him crowing? Aggressive cockerels can also be a problem.

Cock-a-Doodle – Don't!

You cannot stop a cockerel crowing. It is entirely natural and right he should do so. He needs to declare his presence to other poultry in the area at daybreak and he has done this since the first jungle fowl were hatched. All you can do is to use sensible methods to minimize the crowing.

Calming the Crowing

 Only one: If you keep more than one cockerel, they will compete with each other – the same thing happens if there are two or three cockerels within hearing distance of each other. Keep only one cockerel with your hens, especially if in a semi-urban area or even if you simply have close neighbours.

 Move house: Not you, the cockerel. Look at where your hen house is situated. Could you move it away from neighbours and nearer to your house?

Block the sound: Could you screen the crowing by planting a quick-growing barrier at the end nearest to the neighbours? Can you soundproof the house? If you black out the windows at night, be sure you have not restricted the all-important ventilation.

 Be neighbourly: This is probably more for prevention than cure, but try to involve your neighbours in your poultry-keeping activities by taking round the occasional box of eggs and inviting them to come and see them. It is harder to complain about someone you know well and like.

Extreme Methods of Controlling Crowing

These are to be avoided, especially the ones that actually damage the bird. They include caponizing (giving the bird chemicals to 'neuter' it) – illegal in the UK and many other countries; having a vet operate on the bird's vocal chords – this is just plain wrong; and putting the bird in a box overnight so it cannot get its head up to crow. These are mentioned purely because a well-meaning person might advise you of them.

Aggressive Cockerels

If you keep more than one cockerel, they will inevitably compete, leading to lots more crowing and fighting to see who has the girls. Just don't do it. There are always the exceptions who will live together, but do not take the chance.

Attacking Humans and Pets

Some of the large-fowl cockerels, and not the largest ones but the breeds from the medium-size category, can be very sharp and aggressively attack their owners. One traditional cure for this is to catch the bird when he does it, holding him carefully but securely, and make a big fuss of him. The theory is he will be so embarrassed that he won't try again.

Destined for the Table

If he does persist, then you need to get rid of him. Not only do you not want to take a risk of you or your family being injured, but also you do not want any chicks by him either, in case they grow up to be equally aggressive. Put him on the table, where he belongs!

Most Cockerels Are Charming

It's all back to management. Single cockerels with a group of hens can be charming to watch, as they call the girls overfor a morsel of food. They also tend to keep order within the flock, so potential pecking problems are reduced.

Excessively Amorous

It is, however, nature's instinct for the cockerel, on his release from the house in the morning, to want to mate with as many of his girls as he can. Watch out for over-mating, such as when the back of the neck becomes raw or her back becomes bald. If this happens, you will need to remove him or give the hens more space so they can get away. It is possible to buy 'saddles' for the hens, which fit snugly across their backs to protect them from the cockerel's claws when he mates with them.

Did You Know?

'Treading' is the poultry keeper's word for mating, which accurately represents the act when it takes place.

Natural 'Problems'

There are a few things that happen quite naturally in the yearly cycle of a hen that sometimes alarm new poultry keepers, but they are perfectly normal. Make sure you know about the following situations.

Loss of Feathers — the 'Moult'

An adult hen will go into moult once a year. She needs to replace her feathers and have a break from egg laying to replenish her body. It can be alarming to see for the first time, as she will look very tatty indeed. You may even see her skin underneath.

A Draining Time

At this time, she is very vulnerable, not just because she needs to build up her resources, but also because other hens may well peck her when she is in full moult. Remove her if that is the case. You must keep up the nutrition and continue to feed the balanced layers' pellets, plus you might consider a vitamin supplement in her water. It takes a lot out of her body to grow new feathers and to get her body back into egg-laying condition. Help her by nurturing her at this time.

It's Not Mites!

Do not confuse the moult with a bad attack of external parasites – if she has these, then she will be very dull and soon become ill. In the moult, she may be a little subdued, but will otherwise act normally.

Won't Get off Her Eggs

Not only will she not get off her eggs, but she fluffs up her feathers, cackles loudly at you, almost a chicken scream, and pecks you if you try to move her. She is what is known as a 'broody hen', a hen that has decided to sit to hatch eggs. In some parts of the world, they are called 'cluckers' because that's what they do when you try to shift them.

Preventing Broodiness

Regular collection of eggs is essential but, in many breeds it won't solve the problem, especially the tight-sitting bantams, such as the Silkie. If you really do not want any form of broodiness, choose a hybrid breed.

Stopping the Broodiness

You will need a secure 'broody coop', which protects from the weather but is airy, with bars in the front and a wire floor to allow air to circulate and bring down her temperature. She needs to be fed and watered but, although you will provide bedding, you won't provide anything resembling a nesting area. It is not a fail-safe option – some breeds are so persistent that they will just glue themselves to the floor, in which case you may have to consider providing some fertile eggs for her to hatch.

Broodies will sit for three weeks but, if they are sitting on infertile eggs, they may sit longer to the detriment of their health and even to the point of death. It is important not to allow a broody hen to sit pointlessly for week after week, as it will weaken their system.

During the summer months, if more than one hen becomes broody, there can be competition over whose eggs are whose.

My Hen Won't Lay ...

Throughout this book, we have covered why hens stop laying or reduce their egg production (*see* pages 163 and 210). With correct management, you should not have this problem unless the breed is a seasonal layer.

Possible Causes for Not Laying

- **Incorrect nutrition:** One of the most common. Feed a balanced, bagged ration. A laying hen cannot lay on kitchen scraps alone or meet her potential with only wheat.

- **Breed characteristic:** Be clear when you choose the breed how many eggs you might expect. Many breeds do not lay in the dark days of winter. If you want eggs all year round, choose a hybrid.

- **Illness:** A sick hen cannot lay, so always suspect a medical problem if a good layer ceases. It is most likely to be a parasitic problem such as red mite or worms.

- **Moulting:** As discussed, a hen stops laying when she moults.

- **Stress:** A hen will not lay for a few days if she has been stressed, such as through changing homes, and will stop altogether if she feels threatened, such as being bullied.

- **Not ready:** Although you bought her as point of lay, she may not have quite reached that point, so continue with good nutrition until she does. Check with the supplier if she is a hybrid and has not laid after 22 weeks.

Checklist

- **Management is key**: Good management will prevent most problems.

- **Mites**: Watch out for red mite and have a preventive programme. Scaly leg should always be treated and not accepted.

- **Worms**: Ensure you worm your chickens regularly.

- **Know your enemy**: Protect your poultry from predators. Watch out for domestic danger such as from family pets.

- **Discourage vermin**: Do this through cleanliness and safe storage of food.

- **Behavioural problems**: Act immediately if you discover egg eating or pecking other chickens.

- **Cockerels**: Understand how cockerels behave and do not keep more than one with your hens. Control crowing by sensible management.

- **Natural phases**: Support your hens in their yearly moult and be aware of the broody hen.

- **No eggs**: Chickens not laying is often a result of incorrect management or disease and can be corrected.

Adding
to Your
Chickens

Introducing New Birds

When you want to add to your chickens or if you have to replace some of them, you do not have to breed from scratch – you can bring in birds from outside – but how easy is it to introduce new hens to your existing flock? Read on to discover the challenges and solutions.

Things to Bear in Mind

Remember the Pecking Order

New introductions to the group will not be given a warm welcome. Instead, they will be pecked, bullied and generally made to understand that their place is right at the bottom of the order. If you have to add hens, then always add more than one and carefully monitor them so that you can remove the new hens if the skirmish for the pecking order becomes a battle that will wound them. Do not just leave them to suffer.

Add New Hens at Night

Put the new hens in when the birds are roosting and are quiet. There is just an outside chance that the existing flock might accept them more readily.

Do Not Mix Ages

Never put young, immature birds in with older birds – generally, younger birds are more vulnerable for various reasons:

 Bullying and stress: Younger birds will be bullied by older ones. On the other hand, a large number of fit, young, larger birds may cause stress to a smaller number of aged, smaller chickens.

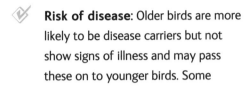

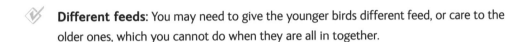

Risk of disease: Older birds are more likely to be disease carriers but not show signs of illness and may pass these on to younger birds. Some diseases will affect younger birds more than they would older birds.

Different feeds: You may need to give the younger birds different feed, or care to the older ones, which you cannot do when they are all in together.

Cockerel: Young birds should not be mated too early – difficult to avoid if you have a cockerel.

Alternative Methods of Introducing Poultry

The Old Switcheroo

One way of unsettling the existing pecking order is to put the new hens in a separate house and then, when they are feeling at home, put the old hens in the new house so that the new hens have the 'upper beak' by already knowing their home. For example, if you introduce, say, six older hens to two new hens, two or three at a time, they may settle down as a group more quickly. There is no guarantee though, so keep watching out for excessive bullying.

Slowly Getting to Know Each Other

This method can work quite well. Place the new hens in a temporary house (but it must be fox- and predator-proof) next to the old house containing the existing hens. After a few days, they will have got used to each other through the safety of a separation barrier. Start letting them out, when you are around to supervise, for an hour or so before they roost, and then finally remove the temporary house and put them in together (the new hens may try go back to their own house – even if you remove it, they will sit where it was).

Breeding Your Birds

One way to increase your flock is to breed your own replacements. This can be done with a broody hen (not necessarily the one who laid the eggs) or with an incubator.

Think About the Boys First

In an average hatch, 50 per cent of the chicks will be male. There is no way you are going to be able to keep this many or, unless they are an extremely rare breed or superb examples, be able to sell them. You will have to rear them to full-grown chickens (for the table) and then either humanely kill them yourself or get someone else to do it. If you really cannot face this, then do not breed chickens, as you will always have to face the fact that there will be surplus males and it will be distressing. See page 184 for discussion of humane slaughter.

How Do Chickens Reproduce?

You will need a cockerel for the eggs to be fertile. Introduce him to the girls, but watch out for any serious aggression either from him or from them. Older hens are not always open to the cockerel's advances and can seriously bully a younger male, especially if he is of a breed that is smaller than them. A single hen with a cockerel can get over-mated, so watch out too for this happening – remove her if she is stressed or in danger of injury. Fit a saddle (see page 223) if she is losing feathers from him treading her.

Wooing the Hen

Prior to the actual act of mating, the cockerel will often 'court' the hen, calling her to him for morsels of food and then dropping one wing and circling her. She is not always impressed by this behaviour!

Cockerels are extremely keen to mate as often as they can and will go to great lengths to do so, from courting the hen to simply chasing her and making a grab for her. It is certainly the first thing on their minds when you let them out in the mornings and can lead to some challenging questions from young children!

The Act of Mating

The cockerel will mount the hen and she will squat down for him. He will 'tread' her, during which the 'cloacal kiss' will take place. The cloaca is a posterior opening that is also known as the vent (*see* page 125). The cloacae touch for only a few seconds, which is long enough for sperm to be transferred from the cockerel to the hen. The cockerel's sperm is within his body (ducks are different, for example, as the male has a penis). The hen is then able to 'store' the semen for up to two weeks.

So How Does the Fertilized Egg Become a Chick?

If the hen is from a breed that is known as a 'sitting' breed (read up on the broodiness of different breeds on pages 78–108), she will lay between 7 and 13 eggs into a nest. If you do not collect them, they will become what is known as a 'clutch' of eggs and she will then become broody. She will sit on her eggs, fluff up her feathers, cackle and even peck if you try to take her off them.

The Developing Unborn Chick

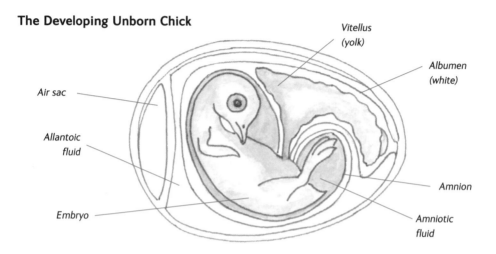

Vitellus (yolk)

Albumen (white)

Air sac

Allantoic fluid

Embryo

Amnion

Amniotic fluid

Egg Development

The fertilized egg will not develop if the hen does not sit on it or if it is not placed in an incubator in order to get a source of heat. This is why eggs are safe to eat even if they are fertilized, as long as you collect them regularly and keep them in a cool environment. The principle behind this is that the egg will stay fertilized but not start to develop until the hen has enough to make a viable brood and then they will all hatch at around the same time. Nature is wonderful!

The Hatching Egg

A hatching egg is a fertilized egg that is intended to be hatched either by a broody hen or by an artificial incubator.

The Natural Mother, Another Hen or Incubation?

If you want the hen who laid the eggs to incubate them and she is of a breed that will do so, then leave her to lay a clutch and, with luck, she should sit on them when she feels she has a sufficient number. If you want to put them in an incubator or under another hen from a broody breed, such as a Silkie or crossbred Silkie, then you will need to get hold of fertilized eggs for hatching.

Did You Know?

A hen's egg takes 21 days to develop from the day that it is incubated, though a bantam may vary from 19 to 21 days.

Obtaining Hatching Eggs

These can be fertilized eggs from your own flock or you can buy hatching eggs from breeders, from a sale or even from the internet. Most people are honest about the breed and fertility of their eggs, but there are some who sell crossbreeds or even eggs of suspect fertility. When an egg does not hatch, it is often difficult to tell whether it's the fault of the original egg or whether something went wrong with the incubation process, so you cannot be sure it was the supplier of the eggs. However, a good supplier of hatching eggs will always be pleased to offer help and advice, so go back to them and share your experiences.

Collection and Storage of Hatching Eggs

Eggs intended for hatching must be collected from chickens who are being mated by a cockerel – you will notice regular matings. They should be collected daily – twice daily, if possible – and stored point-down in a cool place, but not in a fridge. Your hands must be clean; remember a shell is porous and you do not want to pass organisms to the embryo. They must not be cracked or damaged, so have them in a clean container, lined with a soft, clean cloth, as you collect.

On the Clock

A fertilized egg ready for 'setting' (putting under a hen or in an incubator) loses fertility after seven days, so set your collected eggs within a week if possible.

Be Picky

Although you can only incubate clean, unsoiled eggs, do not wash hatching eggs. Discard any dirty eggs, such as those that have been defecated on, and use only clean ones. You must also

discard any cracked eggs, however tiny the damage, and misshapen eggs, such as round or long and thin. Eggs with shell problems such as thick rings caused during laying, should also be discarded. Thin-shelled eggs might break in the incubator or under the hen and cause contamination to the others, so do not risk those either.

Natural Incubation

Natural incubation is when a hen, not necessarily the one who laid the eggs, sits tight on the eggs until they hatch 21 days later. Many experienced breeders are convinced that this is the most reliable way of hatching eggs and go to great lengths to acquire bantams that are naturally inclined to go broody to sit on the eggs of other breeds. Unlike artificial incubation, there is little chance of human error in this method, but the disadvantage is that there is hen error and sometimes chickens can 'let you down' by getting off eggs or even squashing eggs or chicks. Of course, this can often still be traced to 'human error'.

Care of the Broody Hen

The hen will sit tight for 21 days, as she needs to keep the eggs at a constant temperature and also to turn and care for the eggs. She can leave them for 15–20 minutes to eat and drink, so these two things need to be within easy reach – not next to her, but close by. Most of all, she needs protection from predators – she must not sit outside, as so many seem to want to do, hiding themselves in the border beds of your garden!

Moving a Broody Hen

Not as straightforward as is often thought. If you find a hen sitting on a clutch of eggs in a place where you feel she is at risk from predators, you have two choices: move her to a place of safety, or build some protection around her. If you move her, do so in the dark and to a

fairly small space which can be enlarged during the day. Even so, you do get the odd broody who will get off her eggs, insisting that the only place she can be is the dangerous place you removed her from – in this case, there is nothing you can do but start again, or, if the eggs are not cold, remove them to an incubator. However, most really broody hens will stay with their eggs even if you have moved them.

Top Tips for Broody Hens

Beware mites: Make sure she has been treated for mite. If she is bothered by red mite while sitting, she will be forced to get off her eggs and abandon them or, worse still, the chicks will immediately get mite, which will weaken and kill them.

Nutrition: Feed good-quality food and do offer some corn as well. A vitamin supplement in the water may help.

Leave her be: Do not disturb her or allow lots of people to look at her or she may abandon her eggs.

Keep her clean: Broody hen droppings are the largest, smelliest things you can imagine, so remove them from the nest or house.

Keep her separate: If she is anywhere that other hens can access, it is likely that other eggs will be laid into her nest when she gets off to feed. You will not know then which eggs are on their way to hatching and which are new, plus she will end up with too many to 'cover' (sit on with her body).

Artificial Incubation

This is hatching eggs by means of a heating machine, which gives you greater control over when the eggs are hatched, as you don't have to wait for a broody.

Choosing an Incubator

Size: The size of the machine is the first concern. If you only have a small number of chickens, choose a machine that hatches 12 to 24 eggs. Not only will it be cheaper, but it will also mean that you are not tempted to over-hatch – that is, put more eggs into the incubator than you can cope with rearing or keeping long term. Remember, incubators run better at full capacity, so do not over-buy space for eggs.

 New or second-hand?
Whichever you choose, make sure it has a manual – sometimes, you can get manuals for second-hand ones from the manufacturers online, but check before you buy. A second-hand incubator will need thorough cleaning with a suitable incubator disinfectant. The warmth and moistness necessary for eggs are also ideal for bacteria, so the incubator must always be cleaned after every use.

 Type: Choose an incubator that makes life as easy as possible for you (*see below*).

Types of Incubator

There are choices to be made with the type of incubator, though most beginners will go for a forced-air, fully automatic one.

 Still air: The first kind of incubator, it is a heated, insulated box designed to avoid hot or cold spots.

 Forced air: Also known as 'fan-assisted', this is the same box as the still-air kind, with a fan. The advantage is that air gently circulates around the eggs and you can have more than one layer.

Manual (*see right*): A hen will turn her own eggs to prevent the embryo sticking to the shell, so artificially hatched eggs also need turning at least three times a day up to day 18. If an incubator is manual, this has to be done to each egg by hand.

Semi-automatic: The incubator still does not turn the eggs, but a device allows you to turn them all at once from the outside – which still needs to be done at least three times a day.

Fully automatic (*see right*): The incubator turns the eggs on rollers – the great advantage being that it never gets forgotten and the incubator does not have to be opened.

Hatcher: A built-in hatcher means you put trays in on day 18 for the eggs and they hatch within the machine. Otherwise, eggs have to be transferred to a separate hatcher. Be sure to check which sort of machine it is before you buy. A built-in hatcher is better for beginners.

Care of Eggs in an Incubator

It is important to regularly check the eggs to make sure they are developing correctly. What you do not want in a hot, humid atmosphere is an egg that goes bad because it is unfertilized or the embryo has died.

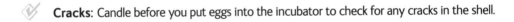

Candling Eggs

Together with your incubator, you should buy a 'candler'. 'Candling' is where you shine a bright light through the egg so that you can see what is happening inside the egg. You can make your own using a torch, but purpose-made ones are quite effective and much cheaper than they were. Candle eggs to check for the following stages:

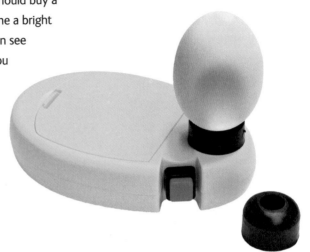

A purpose-made egg candler is preferable to a torch.

- **Cracks**: Candle before you put eggs into the incubator to check for any cracks in the shell.

- **Veins**: At 5–7 days, look to see if red veins are developing. If you can see nothing, the egg is 'clear' and therefore unfertilized. (If unsure, put back in the incubator for a few more days.) Continue to candle weekly and remove any clear eggs, as they may burst and cause contamination to the others.

- **Air sac**: As the embryo develops, the air sac at the broad end (the rounded or 'blunt' end, as opposed to the pointed end) will grow in size – an egg needs to lose around 12 per cent of its weight to allow this to happen. If the air space is too small, the egg is not losing enough moisture, while a very large air space means they have lost too much – both conditions lead to the chick being unable to hatch and therefore found as 'dead in shell'.

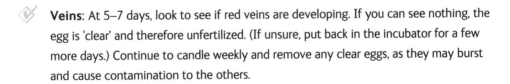

Don't worry too much about being able to recognize immediately if the air sac is the correct size – although if you record what you see in a chart, that will help. Just look for a pattern of development of the air sac to begin with. This is gradual to start with and then rapidly increases towards hatching.

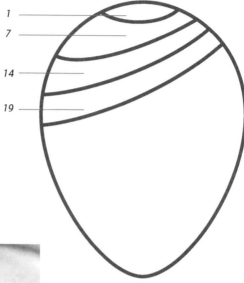

This diagram shows the normal development of the air sac – each number represents the number of days passed.

The Question of Humidity

Getting the humidity right for hatching is something that concerns many people who incubate. For the beginner, the very best thing you can do is to slavishly follow the manufacturer's instructions. They know their machine best and they want it to perform well. If you have a bad hatch, then phone them for advice.

Positioning the Incubator

The incubator should be in a cool, but not cold, place where the temperature remains as even as possible. A sudden, dramatic change in temperature or an extreme of humidity can and will affect the eggs within the machine, sometimes leading to a poor hatching rate.

Keep Records

Doing this will help you to work out what went wrong (and what went right!) in your hatch. You need to monitor the temperature of the incubator, the outside temperature, the weather and the times you added water, plus details of the eggs placed in the machine.

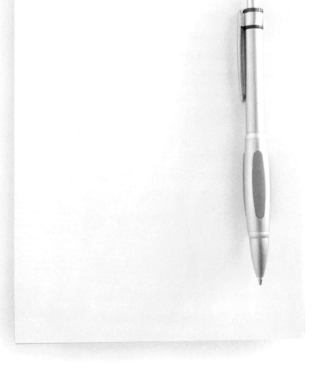

Top Tip

Most incubators run at around 37.5°C (99.5°F), but check your manufacturer's guidelines. Run them and get the temperature established before you set the eggs.

How Eggs Hatch

Eggs hatch the same way whether they are artificially incubated or whether the hen is involved. Around day 18 or 19, the chicks inside the egg begin to use their 'egg tooth' to chip through the egg. This is called 'pipping'. It is very important to leave the hen alone while this is happening (and if in the incubator, leave it

closed) – do not interfere with temperature or humidity at this stage. If you have a broody hen, you may notice that she has begun to cluck very gently to the eggs – the chicks inside can hear this and they respond to her.

Breaking Through

By day 21, they will be ready to hatch completely, having absorbed the remaining egg yolk and will now be working round the broad end of the shell in a ring, ready to push it off with their feet. You must not help them. If you go in too early, then the blood supply will still be connected to the shell membrane and the result will be death for the chick.

Hatched Out

The newly hatched chick is wet but soon dries, whether under the hen or in an incubator. Don't worry about feeding the chick – for at least the first 24 hours, and up to 48 hours, they will be living off the absorbed egg yolk. This is so that the remaining eggs have time to hatch. The broody will be forced to get off the eggs after a couple of days of hatching so that the ones that hatched first can find food and water. The remaining eggs, even if hatching, may

well be left – you can quickly put them in an incubator if you have one. Mostly, the hens have timed it so that the clutch all hatch more or less within 36 hours and the only ones that did not hatch were unfertilized or had other problems. In the incubator, the newly hatched chicks stay for a day in the hatcher part of the machine before being taken out for hand-rearing.

Hen and Chicks

If the eggs have been incubated naturally by the mother, as with her broodiness, your job is to support her, keep her and her new brood safe and to provide suitable feed (chick crumbs for the chicks and some corn for her – she will also eat some of their crumbs) and clean, safe

water (see the rearing section, page 246). Some greenery for her to peck at would be good and, if she is in a run that can be moved, move it so that she has fresh grass on a regular basis. She will do everything, all you need to do is to sit back and enjoy your new family. Remember, she will be very protective, so respect her and do not let family pets near the new brood.

Chick Rearing

As we have seen, the hen will take care of her brood, but for incubator-hatched chicks, it is all down to you. Luckily, given the correct conditions, chicks are generally more than willing to live.

Replacing the Hen

The hen would normally control the warmth of her brood, sitting on them much of the time early on and gradually letting them out of her feathers more frequently until they have a set of their own feathers (at around six weeks) that will help to keep them warm. Even then, very often she will still sit on them in the cool of the night. This is what you have to reproduce for your incubator-hatched chicks.

In addition to a heat source that is completely reliable, the chicks need a safe and draught-free environment, nutritious feed (chick crumbs) and safe water.

Making a Brooder

'Brooders', often known as 'electric hens', provide this environment. You can buy these specially made, which is useful if you are going to raise chicks on a regular basis. Most people, though, make their own.

A Cardboard House

Ideally, construct the brooder from two or three (depending on how big you need it) sheets of hardboard that are about 60 x 180 cm (2 x 6 ft). Bend it to form a circle and use spliced wood or pegs to keep the ends together. This must be in an enclosed building, safe from predatory birds, rats or other predators. If in doubt, net the roof as well, to discourage casual predators. The advantages of this brooder are that:

- **You can adapt its size by bending the hardboard in or pulling it out.**
- **Because it is round, there are no cold corners for chicks to get trapped in.**

If you do not have access to hardboard or you have suddenly to rear young chicks, then use a large box until you can construct a more purpose-made brooder.

The Heat Source

Traditionally, this is a heat-radiant lamp such as an infrared or a ceramic. Choose a dull emitter, as you do not want a bright light shining day and night with your young chicks, as it does not give them sufficient time to rest. Have a spare bulb just in case. You will need to suspend the lamp over your brooder, in the middle.

Don't Cook or Chill Them

Beginners always worry about the correct temperature, which is understandable, but watching your chicks will tell you more than anything. If they are cold, they will huddle together under the lamp and if it is too hot, then they will be desperate to get away from it, panting and looking uncomfortable. Just to give you a start: the ideal temperature is about 34°C

(93°F), which you need to measure at chick height from the floor.

Be Sure

Make sure there is enough space in the brooder for the chicks to be able to get away from the heat source so that, if you have got it wrong, they can seek a cooler place until you sort it out. Chilled chicks will often have sticky bottoms after being cold, so you may need to gently wipe their tiny vents clean.

What to Use as Bedding

Non-dusty wood shavings are the preferred option for most poultry keepers, but any bedding where they can stand and develop strong legs, that is dry and non-dusty, could be considered, such as some of the specialist horse beddings now on the market.

Top Tip

When you first put the chicks in the brooder, check them, check them and check them again. Get up in the night and check them for the first few nights. It is worth it. Your observation is the skill necessary for them to live.

Safe Water

Choose a chick drinker, which is designed to offer only a small area of water – that is, sufficient for their small beaks to access, but not large enough for them to accidentally get into the water. Keep it very clean and available at all times. As the chicks grow, you can raise the water container so it does not become full of bedding, but be sure they can reach it. You can change it to a larger container as they continue to grow. If you have more than a few chicks, use two chick drinkers.

Alternative

If caught out with no chick drinker, use a shallow dish with large (clean) pebbles in the middle and get a proper chick drinker as soon as possible. You can gently dip their beaks in the water briefly to encourage them to drink.

Nutritious Feed

You will have already bought your chick crumbs ready for the hatch! These are designed for their age and are the high in protein and other nutrients, plus they are right size for them to want to peck. I often encourage them to start with, by tapping the crumbs with my finger and making clucking noises; I only hope someone is not watching at the time.

As They Grow ...

When to Turn Off the Heat

It really depends on the time of year as to when you take them off the heat, but you can gradually raise the lamp over the weeks (but not in the first week or so). At six weeks, they will be better-feathered, but some breeds take longer and, of course, if the weather is cold or the temperature drops at night, then you will have to continue longer with the heat. Do not just suddenly remove it.

Turn it off for the warmer hours during the day (or in summer, you might be able to put them outside in a safe house for the daytime) and then turn the lamp off most of the day, perhaps putting it back at night until finally you turn it off completely. Don't remove the lamp for a few days, but continue to observe and then, if all is well and the weather is not extreme, you can congratulate yourself on successfully having reared your chicks to the state poultry keepers call 'off-heat'.

Young Adolescents

From six weeks onwards, again depending on the weather, you can transfer the young chickens into a snug outside house with a daytime run. They are still very attractive to predator birds as well as mammals, so you may need to cover the run. You can start to let them free range when you are there. If you are considering integrating them with the rest of your flock, then please read the section on introducing new birds.

Getting Free of Mum

If the hen has brooded her chicks, then there will come a time when she just abandons them. There is no definite time for this. Some hens carry on for months with their adult brood still clinging round them, while others dump them as soon as they are feathered. I have noticed that the preferred method for a hen is to roost somewhere up high where the adolescents cannot reach, which leaves them moping around on the ground. Watch out for this, because this is when chickens can get accidentally left out of the house at night and become in danger from predators.

Checklist

- **Plan ahead**: Introduce new chickens with care.

- **Hierarchy**: Remember the pecking order and work with it, and don't mix ages or sizes when introducing new chickens.

- **Observe**: Take time to watch how the new birds interact.

- **Think twice**: Don't breed unless you have space, and remember that 50 per cent of chicks hatched will be cockerels – can you deal with them?

- **Fertilized eggs**: These do not develop unless they are incubated, so can be eaten. Store and collect hatching eggs carefully.

- **The natural way**: A broody hen is the easiest method of hatching eggs. Make sure she is protected from predators and provide food and water.

- **Artificial control**: An incubator will give you more control over time of hatching and allow you to hatch without a hen. Consider what you want from an incubator before buying.

- **Monitoring**: Always candle your eggs when in an incubator, but not too often, and keep detailed records of your incubation.

- **New chicks**: Do not help a chick hatch out unless very experienced. Once hatched, chicks need warmth, nutritious food and clean water.

Further Reading

Barker, A., *Eggs: 150 Fabulous Recipes - The Definitive Cook's Guide to Egg Cooking*, Lorenz Books, 2007.

Biggs, D., Gurdon, M., *Hen and the Art of Chicken Maintenance*, New Holland Publishers Ltd., 2003.

Bland, D., *Practical Poultry Keeping*, Crowood Press, 1996

Bridgewater, A., *The Self-sufficiency Handbook*, New Holland Publishers Ltd., 2007.

Constantino, M., *Living Green*, Flame Tree Publishing, 2009.

Damerow, G., *The Chicken Health Handbook*, Storey Books, 2008.

Eastoe. J., *Hen Keeping: Inspiration and Practical Advice for Would-be Smallholders*, Collins & Brown, 2007.

Ekarius, C., *Storey's Illustrated Guide to Poultry Breeds*, Storey Books, 2007.

Goodchild, C., Thompson, A., *Keeping Poultry and Rabbits on Scraps: A Penguin Handbook*, Penguin, 2008.

Gow McDilda, D., *365 Ways to Live Green*, Adams Media, 2008.

Hollander, J., *Chicken Coops for the Soul: A Henkeeper's Story*, Guardian Books, 2010.

Peacock, P., *The Urban Hen: A Practical Guide to Keeping Poultry in a Town or City*, Spring Hill, 2009.

Laughton, R., *Surviving and Thriving on the Land: How to use your time and energy to run a successful smallholding*, Green Books, 2008.

Popescu, C., *Chicken Runs and Vegetable Plots*, Cavalier Paperbacks, 2009.

Popescu, C., *Hens in the Garden, Eggs in the Kitchen*, Cavalier Paperbacks, 2003.

Roberts, M., *Poultry House Construction*, Gold Cockerel Books, 1997.

Roberts, V., *British Poultry Standards*, Wiley-Blackwell, 2008.

Roberts, V., *Diseases of Free-Range Poultry: Including Ducks, Geese, Turkeys, Pheasants, Guinea Fowl, Quail and Wild Waterfowl*, Whittet Books Ltd., 2009.

Seymour, J., *The New Complete Book of Self-Sufficiency: The classic guide for realists and dreamers*, Dorling Kindersley, 2003.

Scrivener, D., *Starting with Bantams*, Broad Leys Publishing Limited, 2002.

Thear, K., *Incubation: A Guide to Hatching and Rearing*, Broad Leys Publishing Limited, 1997.

Thear, K., *Keeping Quail: A Guide to Domestic and Commercial Management*, Broad Leys Publishing Limited, 2005.

Verhoef-Verhallen, E.J.J., Rijis, A., *Complete Encyclopedia of Chickens*, Rebo Productions, 2004.

Websites

www.avianweb.com
Has information for breeders and owners that includes a wide range of health information.

www.backyardchickens.com
Help for those curious about raising chickens in their own back yard.

www.brinsea.com
Brinsea is the leading manufacturer of egg incubators, hatchers and brooders. The website also provides helpful advice on incubation topics.

www.chickenrecipes.org
A massive variety of recipes for chicken dishes.

www.cotswoldchickens.com
Sells both chickens and chicken housing as well as a help section for new chicken owners.

www.eggrecipes.co.uk
Recipes that require eggs for quick, low-calorie, family or vegetarian meals.

www.freerangepoultry.co.uk
Offers information regarding free-range chicken raising.

www.hsa.org.uk
The The Humane Slaughter Association works towards the highest worldwide standards of welfare for food animals. Provides advice on how to go about humane slaughter.

www.keeping-chickens.co.uk
A blog site that details one family's experiences and advice while raising chickens.

www.omlet.co.uk
Sells modern and inventive housing, equipment and feed, as well as the animals themselves, from chickens to rabbits to bees. It also has a section that offers advice on raising the various animals it sells.

www.pandtpoultry.co.uk
P and T Poultry specialize in supplying a broad selection of chicken supplies, from coops to incubators.

www.poultry.allotment.org.uk
Contains a wide variety of information on keeping many types of chickens, geese and ducks.

www.poultryclub.org
The website for The Poultry Club of Great Britain, an organization that supports the preservation of pure-breed poultry; contains information about egg production, in-depth detail of poultry breeds and information about poultry shows.

www.poultrypages.com
Has information on breeding and raising poultry, as well as the supplies necessary to do so.

www.self-sufficient.co.uk
Offers advice for those who are looking to become more self-sufficient and for smallholders, including chicken owners.

www.smallholder.co.uk
The website for *Smallholder* magazine, offering news and advice on all things helpful to the back-yard self-sufficiency convert as well as smallholders – from poultry and livestock keeping to growing your own food.

www.sprcentre.co.uk
The only site offering free person-to-person advice from an experienced trained poultry specialist.

Index